CASE STUDIES IN FORENSIC EPIDEMIOLOGY

CASE STUDIES IN FORENSIC EPIDEMIOLOGY

Sana Loue
Case Western Reserve University
Cleveland, Ohio

Springer Science+Business Media, LLC

Library of Congress Cataloging-in-Publication Data

Loue, Sana.
Case studies in forensic epidemiology/Sana Loue.
p. cm.
Includes bibliographical references and index.

DOI 10.1007/978-0-306-47524-5
1. Forensic epidemiology—Case studies. I. Title.
[DNLM: 1. Epidemiologic Methods—United States—Legal Cases. 2. Epidemiology—legislation & jurisprudence—United States—Legal Cases. 3. Expert Testimony—United States—Legal Cases. 4. Liability, Legal—United States—Legal Cases. 5. Public Health—legislation & jurisprudence—United States—Legal Cases. WA 33 AA1 L886c 2002]
RA1165 .L678 2002
614'.1—dc21

2002016195

Originally published by Kluwer Academic/Plenum Publishers, New York in 2002
MyCopy version of the original edition 2002

http://www.kluweronline.com

10 9 8 7 6 5 4 3 2 1

A C.I.P. record for this book is available from the Library of Congress

Tous ceux dont la vie se passe á chercher la vérité savent bien que les images qu'ils en saisissent sont nécessairement fugitives. Elles brillent un instant pour faire place à des clarets nouvelles et toujours éblouissantes. Bien différente de celle de l'artiste, l'oeuvre du savant est fatalement provisoire. Il le sait et s'en réjouit, puisque la rapide vieillesse de ses livres est la preuves même du progrés de la science.

H. Pirenne, quoted in G. Gérardy (1962). *Henri Pirenne, 1862–1935*. Brussels: Ministère de l'education nationale et de la culture, Administration des services educatifs.

All those whose lives are spent searching for truth are well aware that the glimpses they catch of it are necessarily fleeting, glittering for an instant only to make way for new and still more dazzling insights. The scholar's work, in marked contrast to the artist, is inevitably provisional. He knows this and rejoices in it, for the rapid obsolescence of his books is the very proof of the progress of scholarship.

J. Boswell (1980). *Christianity, Social Tolerance, and Homosexuality: Gay People in Western Europe from the Beginning of the Christian Era to the Fourteenth Century.* Chicago: University of Chicago (translating H. Pirenne, above).

Preface

Epidemiology has often been defined as the study of the distribution of disease, together with the distribution of factors that may modify that risk of disease. As such, epidemiology has often been reduced to a methodology only, providing a mechanism for the study of disease that is somehow removed, separate and apart from the populations that serve as its focus.

Epidemiology, however, is much more than that. The discipline provides a way of perceiving and knowing the world, and of relating to the communities whose health and disease patterns we are trying to understand. As such, its usefulness extends past the construction of questionnaires, the detective work inherent in tracing the source of an infection or the analysis of data. Rather, epidemiology serves as a point of reference and a linkage between various domains of reality: in the courtroom, between a community's injuries and those alleged to be responsible for those violations; between the community striving to effectuate changes to improve its health and environment and the lawmakers and policymakers whose actions may dictate or control the likelihood of that change; and between "mainstream" populations and those who become or remain marginalized and stigmatized due to disease or perceived disease.

Many epidemiologists may balk at this use of their profession, arguing that, at best, it is an unnecessary foray from the protected, objective world of science into the relatively subjective and highly charged world of public affairs and legal wars. What this view ignores, however, is our own subjectivity as epidemiologists, inherent in our everyday decisions relating to the value of what it is we are to study, the relative importance of the research questions that we wish to tackle. Those decisions are not merely a function of calculating the monetary balance between monies to be expended and costs saved through prevented cases of disease. We make, in many ways, the value judgments that we profess to avoid.

This text focuses on various uses of epidemiology in the legal context, as it is broadly defined. Chapter 1 focuses on forensic epidemiology as it is usually conceived of. Concepts of causation in law and in epidemiology are explored, as are the mechanisms for determining causation. Chapters 2 and 3 expand the concepts discussed in chapter 1 by providing case studies of the silicone breast implant litigation and of a lawsuit involving exposure to *E. coli.* Chapters 4 through 6 continue with an examination of the uses of epidemiology in the legislative and regulatory arenas by providing a foundation, followed by a discussion of FDA action related to silicone breast implants and tobacco.

Chapters 7 through 12 depart from what is often traditionally thought of as the legal forum. These chapters focus on the use of epidemiology in the context of community advocacy and social change. Readers may wonder at the inclusion of such topics and their relation to law and epidemiology. Community advocacy efforts often utilize legal mechanisms, among others, to accomplish their aims. Epidemiology may be used as a basis for these actions. Consider, for instance, the efforts of activists to establish needle exchange programs, revisions in FDA drug approval processes, and the control of tobacco. As society changes, our perceptions of what is harmful or healthful, normal or deviant, changes, as well as our knowledge and understanding evolve. We see this, for example, in our current reliance on "risk behaviors," in lieu of "risk groups," to designate those individuals who are at greater risk of contracting HIV infection, and in the evolution of attitudes regarding the medical use of marijuana.

This text, above all else, challenges the reader to push the boundaries of his or her thinking about what epidemiology is and how it is to be used to make a contribution. The reader is asked to arrive at a more complete understanding of the populations with whom we work, the diseases that we investigate, and the systems that underlie and shape our endeavors.

Acknowledgments

The author gratefully acknowledges the helpful review and critique of earlier versions of this manuscript by Dr. Siran Koroukian. Stephanie Stewart and Gary W. Edmunds deserve praise and thanks for their time and diligence in locating needed materials. Dr. Hal Morgenstern has generously permitted the use of his diagrams illustrating the finer points of confounding. The excerpts from various trials and depositions would not have been available but for the interest and generosity of several attorneys and their law firms: the law office of Jerry Moberg in Euphrata, Washington, Jerome L. Skinner of the Cincinnati, Ohio law firm of Waite, Schneider, Bayless, and Chesley, and Dianna L. Pendleton. Mariclaire Cloutier, the editor of this volume, deserves praise for her patience and support.

Contents

CASE STUDIES IN FORENSIC EPIDEMIOLOGY

CHAPTER 1

Epidemiology in the Courtroom

Dissonant Goals, Divergent Processes

> The task before us is more daunting still when the dispute concerns matters at the very cutting edge of scientific research, where fact meets theory and certainty dissolves into probability. As the record in this case illustrates, scientists often have vigorous and sincere disagreements as to what research methodology is proper, what should be accepted as sufficient proof for the existence of a "fact," and whether information derived by a particular method can tell us anything useful about the subject under study (Judge Alex Kozinski, *Daubert v. Merrell-Dow Pharmaceuticals, Inc.*, 1995: 43 F.3d at 1316).

EPIDEMIOLOGY AND LAW: DISSONANT GOALS

With increasing frequency, policymakers, judges, lawyers, and juries are relying on epidemiological findings as a basis for regulatory action, for guidance in the resolution of civil lawsuits for damages claimed in connection with injuries alleged to have occurred as the result of specified exposures, and for the determination of culpability in criminal prosecutions. For instance, the Food and Drug Administration (FDA) relied on epidemiological studies in requiring that manufacturers place a warning label on aspirin bottles to inform potential consumers of the association between Reyes syndrome and aspirin use (Novick, 1987; Schwartz, 1988). Epidemiology was in the forefront in litigation involving numerous exposures, such as Agent Orange, Bendectin, tampons, silicone breast implants, and tobacco (*In re Agent Orange Products Liability Litigation*, 1984; *In re Bendectin Litigation*, 1988; *Kehm v. Procter and Gamble Manufacturing Company*, 1983; Fitzpatrick and Shainwald, 1996; Glantz, Slade, Bero, Hanauer, and Barnes, 1996; Sanders, 1992; Schuck, 1987). And, knowledge of the epidemiology of HIV infection has provided the basis for the formulation and passage of statutes criminalizing sexual relations between HIV-infected individuals and others without disclosure of the individual's HIV seropositivity.

However, this ever-increasing reliance on epidemiological findings as one of the bases for the development of relevant policy and the resolution of disputes and harms has not been uncontroversial. Writers have attacked courts and attorneys for what they see as a blatant disregard of scientific objectivity in favor of easier solutions and the satisfaction of baser human desires (Angell, 1996). Legal scholars have countered these charges with their own analyses of the weaknesses inherent in the medical and legal service delivery systems (Dresser, Wagner, and Giannelli, 1997). Accordingly, an understanding of the different goals between epidemiology and law is critical to an evaluation of whether and how epidemiological findings should be utilized in the context of litigation, rulemaking, or policy development. Justice Harry Blackmun astutely observed:

> It is true that open debate is an essential part of both legal and scientific analyses. Yet there are important differences between the quest for truth in the courtroom and the quest for truth in the laboratory. Scientific conclusions are subject to perpetual revision. Law, on the other hand, must resolve disputes finally and quickly (Daubert, 1993: 113 S.Ct. at 2798).

This requirement, and need, for the final resolution of disputes is reflected in various legal principles. Individuals' ability to bring legal action against another party for an alleged wrong is confined to a specific time frame by the imposition of statutes of limitations; if a complaint is not initiated within a prescribed period of time, the ability to sue and recover damages is forever lost. It is irrelevant to the application of the statute of limitations that at the time the period is to expire or, in legal parlance, the statute of limitations is to run out, the state of scientific knowledge with respect to the alleged cause of the alleged injury is in flux. The legal concept of *res judicata* means that once a final decision has been rendered by the courts in a particular matter, the case cannot be re-instituted again (Marcus and Rowe, Jr., 1994). Suppose for instance, that an individual claims that he contracted a particular form of cancer due to exposure to a specific substance with which he came into regular contact because it was dumped into a nearby river by a polluting corporation. The original jury finds, based on all the scientific evidence that is presented both for and against his claim, that there is no link between the pollution and his cancer and denies an award. This judgment is affirmed on appeal. Five years later, the government issues a report indicating that there is, indeed, a link between this particular substance and certain types of cancer. The gentleman cannot now go back into court, claiming that he should be able to retry his case because the state of scientific knowledge has improved. He has already had his day in court.

Many scientists have implicitly, if not explicitly, recognized the dissonance in purpose between the two domains. Huber, for instance, has characterized science as "a search for absolute and immutable truths. The search does progress. But it does not end" (Huber, 1991: 214). As a consequence, several scientists have recommended, more or less vigorously, the separation of the two spheres. Rothman, for instance, has explained that

> since all scientific theories could be wrong, policy makers should weigh the consequences of actions under various theories. Scientists should inform policy makers about scientific theories, and leave the choice of a theory and an action to policy makers. Not many public health scientists are inclined towards such a strict separation between science and policy, but as a working philosophy it has the advantage of not putting scientists in the awkward position of being advocates for a particular theory .... Indeed, history shows that skepticism is preferable in science (Rothman, 1986: 20).

Poole has adopted a similar position:

> Although science and decision making are both important, they are not the same. In science, we seek to learn, to explain, and to understand. In decision making, we seek reasons to act or to refrain from acting. It demeans neither of these enterprises to acknowledge that they are different from each other. Consider on the one hand the theory that contraceptive diaphragms cause urinary tract infections and, on the other, decisions about the use of diaphragms. The causal theory is either true or false. We can criticize and empirically test it, but we cannot decide that it is or is not true. Its truth or falsity is completely independent of any decision we might make about it. What we can decide is what our own actions, both personal and public, will be with respect to the use of diaphragms (Poole, 1987: 195).

Just as the disciplines of law and science define "Truth" differently, so, too, do they utilize different procedures for the determination of that "Truth," used here as a synonym for causation.

EPIDEMIOLOGY AND LAW: DIVERGENT PROCESSES

Causation in Epidemiology

Background

Epidemiology seeks to answer such questions as "Why do some people contract certain illnesses more frequently than others?" and "Why does a specific illness progress more rapidly in some people compared to others?" The simplest approach to causality in such circumstances is that of pure determinism.

Pure determinism posits specificity of cause and specificity of effect, *i.e.*, that the factor being examined as a cause is the one and only cause of the disease under examination, and that the disease under investigation is the only effect of that factor (Kleinbaum, Kupper, & Morgenstern, 1982). This implies that the factor under examination is both a necessary and sufficient cause of that disease.

Robert Koch's formulation of the criteria for disease causality, in essence, operationalized pure determinism (Kleinbaum, Kupper, & Morgenstern, 1982). Koch's work on tuberculosis provided the basis for his refinements of causation criteria to include five elements:

T1. An alien structure must be exhibited in all cases of the disease.
T2. The structure must be shown to be a living organism and must be distinguishable from all other micro-organisms.
T3. The distinction of micro-organisms must correlate with and explain the disease phenomena.
T4. The micro-organism must be cultivated outside the diseased animal and isolated from all disease products which could be causally significant.
T5. The pure isolated micro-organism must be inoculated into test animals and these animals must then display the same symptoms as the original diseased animal (Carter, 1985).

Koch's causational model has been criticized for its various limitations, including its failure to recognize (1) the multifactorial etiology of many diseases; (2) the multiplicity of effects associated with specific factors; (3) the complexity of many causal factors; (4) our incomplete understanding of disease and disease processes; and (5) the limitations inherent in our ability to measure the causal process (Kleinbaum, Kupper, & Morgenstern, 1982).

Modified determinism addresses the limitations of Koch's model. Rothman explains this model as follows:

> A *cause* is an act or event or a state of nature which initiates or permits, alone or in conjunction with other causes, a sequence of events resulting in an *effect*. A cause which inevitably produces the effect is *sufficient* ... .
>
> * * * *
>
> A specific effect may result from a variety of different sufficient causes ... . If there exists a component cause which is a member of every sufficient cause, such a component is termed a *necessary* cause ... . (Rothman, 1976).

Hence, this model recognizes both that a cluster of factors, rather than a single agent, may produce an effect, and that a specific effect may be the product of various causes. The strength of a specific causal factor depends upon the relative prevalence of component causes. A factor, even though rare, may constitute a strong cause if its complementary causes are common (Rothman, 1986). Two component causes of a sufficient cause are said to be synergistic, in that their joint effect exceeds the sum of their separate effects (Rothman, 1976). As an example, individuals exposed to asbestos are at increased risk of cancer if they also smoke (Hammond, Selikoff, & Seidman, 1979).

This modified deterministic model, however, also has its limitations. We are often unable to identify all of the components of a sufficient cause (Rothman, 1986). Consequently, epidemiologists utilize probability theory and statistical techniques to assess the risk of disease resulting from exposure to hypothesized causal factors (Rothman, 1986). Causation may be inferred by formulating general theories from observation (induction) or by testing general theories against observation (deduction) (Weed, 1986).

Using an inductive approach, Hill has enunciated the criteria to be considered in identifying causal associations: (1) strength; (2) consistency; (3) specificity; (4) temporality; (5) biological gradient; (6) plausibility; (7) coherence; (8) experimental evidence; and (9) analogy (Morabia, 1991). Each of these criteria is discussed briefly below.

The "strength" of an association between the putative causal factor and the effect is dependent on the relative prevalence of other component causes. This criterion encompasses two separate issues: the frequency with which the factor under investigation is found in cases of a specific disease and the frequency with which the factor occurs in the absence of the disease (Sartwell, 1960). "Consistency" refers to the repeated observation of an association between the putative causal factor and the effect in varied populations at varied points in time and under different circumstances (Susser, 1991). Inconsistency, however, does not necessarily negate a causal relationship because all causal components must exist to bring about the effect, and some may be absent. "Specificity" refers to the association between a postulated cause and a single effect (Rothman, 1986). Hill specifically cautioned against overemphasizing the importance of this particular element (Hill, 1965).

"Temporality" requires that the cause precede the effect in time (Rothman, 1986). Although a dose-response curve, or "biological gradient," is to be considered, it does not necessarily indicate causation due to the effects of confounders. "Plausibility" requires that the hypothesized relationship between the causal factor and the effect be biologically plausible. This is clearly limited by the state of our knowledge at any point in time. "Coherence" requires that a postulated causal association be consistent with our knowledge of the natural history and biology of the disease in question. Experimental evidence is rarely available for human populations. "Analogy" posits that reference to known examples, such as a causal association between one drug and birth defects, may provide insights into other causes of birth defects, such as another drug (Rothman, 1986).

Numerous factors must be considered in evaluating whether there exists a causal association between a specific exposure and a specific outcome. These factors include the study design; the measures resulting from the study; the presence, absence, and extent of bias and confounding; the extent to which strategies used to enhance internal validity have been successful; and the ability to assess random error. Each of these elements is discussed below.

An Overview of Research Design

Research design provides a means by which to examine the relation of cause and effect in a specific population (Susser, 1991). This section explores the basic study designs used in epidemiological research, as well as several hybrid designs. Other resources should be consulted for a more extensive discussion.

Epidemiologic research can be classified into three major types: experimental, quasi-experimental, and observational. With each type of research, a variety of different designs can be utilized. Each of these types is described briefly below, followed by a discussion of specific study designs.

Experimental Research

Experimental research often, but not always, involves the randomization of individuals into treatment groups. The treatment groups are also known as study arms, to which individuals are randomly allocated in order to receive a particular treatment under investigation (treatment group) or an alternative treatment or placebo (control group). Randomization is considered essential as a safeguard against selection bias and as insurance against accidental bias (Gore, 1981), *i.e.*, to ensure that both groups are representative of the population as a whole or that both groups are similar in all respects except the treatment that is under study. Experimental studies conducted in the laboratory are usually of relatively short duration and are often used to test etiologic hypotheses, to estimate treatment effects, or to examine the efficacy of an intervention (Kleinbaum, Kupper, & Morgenstern, 1982).

Clinical trials are experimental studies of much longer duration. They have been referred to as "the epidemiologic 'gold standard' for causality inference" (Gray-Donald & Kramer, 1988: 885). They are usually conducted to test the efficacy of a specific intervention, to test etiologic hypotheses, or to estimate long term health effects. Both the experimental group and the control group consist of patients who have already been diagnosed with the disease of interest or are at risk of the disease of interest in a clinical trial involving a prevention. Additionally, the patients must have agreed to be randomized to one of the study arms. Clinical trials often utilize double blinding, whereby neither the individuals conducting the study nor the participants in the study know who is in the treatment group and who is in the control group (Senn, 1991).

Community interventions are also usually of longer duration. They are often initiated to test the efficacy and the effectiveness of a particular health intervention.

Experimental studies offer numerous advantages, including the ability, through randomization, to control for extraneous factors that may be related to the outcome under examination. Unfortunately, the study population ultimately selected through this process may not be comparable to the target population with respect to important characteristics.

Quasi-Experimental Research

Quasi-experiments involve the comparison of one group to itself or of multiple groups. Although this study design permits the investigator to manipulate the study factor, as in an experiment, randomization is not used. Quasi-experiments are most often conducted in a clinic or laboratory setting to test etiologic hypotheses, to evaluate the efficacy of an intervention, or to estimate the long term health effects of an intervention. Those conducted in the program and policy arenas are often devoted to the evaluation of programs or interventions or to an analysis of the costs and benefits of an intervention (Kleinbaum, Kupper, & Morgenstern, 1982). Quasi-experiments are generally smaller and less expensive than experimental studies. However, the investigator has less control over the influence of extraneous risk factors due to the lack of randomization (Kleinbaum, Kupper, & Morgenstern, 1982).

The following scenario provides an example of a quasi-experimental study. Assume, for instance, that exposure to a particular pesticide is thought to be a potential risk factor for the development of specific respiratory illnesses. Exposure to the pesticide in a particular community is ubiquitous, due to its extensive use in agriculture and the consequent leaching of the substance into the soil and the water table. Residents of the small farming area are offered an opportunity to relocate. About one-half of the residents elect to remain, and the others choose to leave. The members of each group are then followed over time to assess the effect of the continuing exposure or lack thereof on the development of the respiratory illnesses in question. If this had been an experiment, rather than a quasi-experiment, the residents would have been randomized to the exposure. Instead, the residents-participants made their own decisions with respect to the continuing exposure.

Observational Studies

Observational studies are the most frequently utilized type of study in epidemiology. Unlike experimental and quasi-experimental studies, they do not involve the manipulation of the study factor. In some cases, manipulation of the exposure under study might be ethically and/or legally problematic. For instance, exposing study participants to a suspected carcinogen in order to confirm or refute the substance's association with cancer would be unethical because participants would be deliberately exposed to a suspected risk of harm. Logistically such a study would be problematic because of the long follow-up period that would be required in order to detect the increased risk of cancer. The goal of observational studies is to arrive at the same conclusions that would have resulted from an experiment (Gray-Donald & Kramer, 1988). Observational studies may be descriptive or etiologic in nature. Descriptive studies are often used to estimate the frequency of a specific disease in a population or to generate hypotheses or ideas for new interventions. Etiologic studies can be used not only to generate etiologic and preventive hypotheses, but can also be used to test specific hypotheses and estimate health effects. Observational studies afford the investigator much less control over extraneous risk factors than do either experimental or quasi-experimental studies because they are conducted in a natural setting (Kleinbaum, Kupper, & Morgenstern, 1982).

Observational studies can use any of numerous study designs, depending on how the research question is framed, the current state of knowledge with respect to the disease or exposure/risk factor at issue, the costs of the proposed study, and various other considerations. Observational study designs include cohort study design, case-control design, and cross-sectional design. Other common observational designs include the etiologic study, the proportional study, space-time cluster studies, and the family cluster study (Kleinbaum, Kupper, & Morgenstern, 1982).

Clinical trials and observational studies are particularly central to epidemiology. For this reason, clinical trials and various types of observational study designs are addressed in further detail below.

Study Designs

Clinical Trials

Clinical trials of new drugs are conducted in three phases. The first phase consists of the initial introduction of the drug into humans. Phase I studies are conducted to assess the metabolic and pharmacologic actions of the drugs in humans, the side effects of the drug, and the effectiveness of the drug. Phase I studies, which are closely monitored, are generally limited to 20 to 80 participants (21 Code of Federal Regulations section 312.21(a), 1998).

Phase II studies build on the knowledge that was obtained from Phase I studies. Phase II studies are carried out in patients with the disease under study for the purpose of assessing the drug's effectiveness for a particular indication and for determining the drug's short-term side effects and risks. Typically, Phase II studies involve up to several hundred participants (21 Code of Federal Regulations section 312.21b, 1998).

Phase III studies incorporate the knowledge gained through Phase I and Phase II trials. Phase III trials focus on gathering additional data relating to the drug's effectiveness and safety. Phase III studies are potentially quite large, sometimes involving thousands of participants (21 Code of Federal Regulations section 312.21c, 1998).

The initial phase of planning a clinical trial requires that a decision be made regarding the treatment(s) to be studied and the eligibility criteria for individuals to participate in the trial (Rosner, 1987). Clinical trials designed to answer questions relating to biological response generally enroll a rather homogenous participant population, in order to reduce the variability between participants and simplify the analysis of the results. Patients enrolled in such a study must be sufficiently healthy so that they do not die before the end of the study, but not so well that they recover from the disease.

Larger clinical trials, particularly multisite trials, often enroll more heterogeneous participants. This situation more closely mirrors everyday clinical practice with diverse patients and also comports with ethical considerations and federal regulations regarding the equitable distribution of the benefits and the burdens of research.

During a phase III clinical trial, the treatment under study is compared to one or more standard treatments, if a treatment exists, or to a placebo.

A placebo has been defined as

> Any therapy or that component of any therapy that is intentionally or knowingly used for its nonspecific, psychological, or psychophysiological therapeutic effect, or that is used for a presumed specific therapeutic effect on a patient, symptom, or illness but is without specific activity for the condition being treated ... [and], when used as a control in experimental studies, is a substance or procedure that is without specific activity for the condition being treated. (Shapiro and Shapiro, 1997: 41)

Participants are randomized to one of the treatment arms or, if a placebo arm is used, to the treatment arm(s) or placebo arm. The "gold standard" for clinical trials requires that neither the researcher(s) nor the participants know until the conclusion of the study which individuals are receiving the experimental treatment and which are receiving the standard treatment or placebo, i.e. the study is double-blinded.

Additionally, researchers must decide which endpoint(s) to use as a basis for evaluating participants' response to the treatment (O'Brien & Shampo, 1988). The endpoint(s) will vary depending on the disease and the treatment under investigation, but may include recurrence of an event, such as myocardial infarction; quality of life; functional capacity; or death. Clinical trials are often concerned not only with whether an endpoint occurs, but also with the length of time until its occurrence. As an example, a clinical trial for a cancer treatment may be concerned with not only the recurrence of the malignancy, but also with the time between the treatment and the reappearance of the disease.

Ethical concerns have been raised about classic clinical trials on a number of grounds. Clinicians may believe that patients should not be randomized because it may deprive them of an opportunity to receive a new alternative drug (Farrar, 1991). Others believe that randomization is inappropriate in situations in which the patient has exhausted all available therapies or one in which the standard therapy has provided no benefit (Rosner, 1987).

Crossover designs have also been proposed as an alternative to classic clinical trials and as a mechanism for addressing variations between patients in response to treatment. With a crossover design, half of the patients are randomized to Group 1 and the second half to Group 2. Following administration of Treatment A to Group 1 and Treatment B to Group 2, the allocation of treatment is reversed. Generally, crossover designs incorporate an appropriate "washout period" between administration of the treatment to each group, in order to reduce the possibility of a carryover effect from the first treatment period to the second (Hills & Armitage, 1979). Crossover designs are most useful in situations in which the treatment under investigation is for the alleviation of a condition, rather than the effectuation of a cure.

Observational Designs

There are many observational study designs that are potentially available to the researcher, depending upon the research question to be posed and the extent of the funding available to conduct a particular study. Some of these designs are hybrids of the more basic designs that are discussed here: cohort studies, case-control studies, cross-sectional studies, and ecological study designs.

Cohort Studies. Cohort studies can be conducted prospectively or retrospectively. Prospective cohort studies require the identification and classification of initially disease-free individuals into categories according to whether they have or have not been exposed to the factor under study. Each group is followed over time in order to observe the number of new cases of the disease under investigation that occurs in each group in a specified period of time (Kelsey, Thompson, & Evans, 1986).

Numerous difficulties may attend prospective studies. First, individuals may already have been exposed to the factor under study and the length and intensity of that exposure may be difficult to ascertain. As an example, a cohort study to examine the effects of an occupational exposure would consist of a group exposed to the substance under study and a group that was not exposed. Depending upon the particular industry and configuration of the workplaces, however, some members of the group classified as unexposed may, in fact, have been exposed to small amounts of the substance. Second, although individuals enrolled into the study may be believed to be free of the disease, the disease process may, in fact, have commenced in some but may be undetectable by diagnostic methods and tools then available. This could be true, for instance, in studies involving cancer or schizophrenia. Third, a minimum length of time following exposure may be required to allow a biologically appropriate induction time, as well as a subsequent period of time after causation but before disease detection (latent period). In situations where we do not have complete knowledge of the induction and latency periods, we must make assumptions about the lengths of these times (Rothman, 1986). Fourth, prospective cohorts also require large sample sizes and are often quite costly (Kelsey, Thompson, & Evans, 1986). (What is considered a "large sample size" varies depending upon the disease under study, the exposure under study, and various other factors.) Fifth, the choice of a comparison group of unexposed individuals may be quite difficult. Too, we tend to think of disease as being present or absent, but some diseases, such as high blood pressure, may occur along a spectrum, making classification of individuals as diseased or not diseased more complex (Kelsey, Thompson, & Evans, 1986).

Individuals in a prospective cohort study are to be followed over time. However, some individuals may drop out of the study or be lost to follow-up. These losses may be related to disease status, thereby producing a bias in the measurement of disease (Kelsey, Thompson, & Evans, 1986). It is easy to imagine that as someone becomes progressively more ill, that he or she may not want to undergo a physical examination or respond to questions relating to the illness. Information on other extraneous variables that may affect the results may not be available (Kelsey, Thompson, & Evans, 1986).

Retrospective cohort studies, also known as historical cohort studies, require the identification of individuals based on their past exposure and the reconstruction of their disease experience up to a defined point in time. Retrospective cohort designs are often useful for examining the effects of occupational exposures. Unlike prospective cohort studies, they often rely on already-existing records and may consequently be completed in less time and with lesser cost than a prospective cohort study. Retrospective cohort studies do, however, share some of the same difficulties as prospective studies, including difficulties in the ascertainment and measurement of extraneous relevant characteristics (confounding variables), and difficulties tracing individuals through time. Despite the difficulties inherent in cohort designs, cohort studies offer a major benefit: the ability to calculate incidence rates for the exposed and the unexposed groups. As an example, a retrospective cohort study examining the relationship between exposure to silicone breast implants and the development of a specific autoimmune disorder might follow a cohort of women from a point in time 20 years ago to determine how many cases of the specific autoimmune disorder developed in those who did and did not have silicone breast implants.

Case-control Studies. Unlike cohort studies which follow individuals through time after classifying them based on their exposure status, case-control studies require the classification of individuals on the basis of their current disease status and then examine their past exposure to the factor of interest. Case-control design has been used to investigate disease outbreaks (Dwyer, Strickler, Goodman, & Armenian, 1994); to identify occupational risk factors (Checkoway & Demers, 1994); in genetic epidemiologic studies (Khoury & Beaty, 1994); for indirect estimation in demography (Khlat, 1994); to evaluate vaccination effectiveness and vaccine efficacy (Comstock, 1994); to evaluate treatment and program efficacy (Selby, 1994); and to evaluate the efficacy of screening tests (Weiss, 1994). Case-control studies are most useful in evaluating risk factors for rare diseases and for diseases of rapid onset. With diseases of slow onset, it may be difficult to ascertain whether a particular factor contributed to disease causation or arose after the commencement of the disease process (Kelsey, Thompson, & Evans, 1986).

The conduct of a case-control study requires selection of the cases (the diseased group) and the controls (the nondiseased group) from separate populations. It should be obvious that prior to selecting the cases, the investigator must define what constitutes a case conceptually. This is not as simple a task as it first appears, particularly when the disease condition is a new and relatively unstudied entity (Lasky & Stolley, 1994). For instance, case-control design was utilized in a number of studies that attempted to determine whether an association exists between silicone breast implants and the development of an autoimmune disorder (Chang, 1993; Hirmand, Latenta, and Hoffman, 1993; Swan, 1994). However, the diagnosis of "human adjuvant disease," the term used to refer to the syndrome alleged to have resulted from exposure to silicone, was difficult due to the broad range of symptoms encompassed by that term and the relative lack of research that existed initially. Causal inference is possible only if one assumes that the controls are "representative of the same candidate population ... from which the cases ... developed ... ." (Kleinbaum, Kupper, & Morgenstern, 1982: 68). Consequently, the selection of appropriate cases and controls is crucial to the validity of the study.

Methods and criteria for the selections of appropriate cases and controls have been discussed extensively in the literature and will only be summarized here. Cases are often selected from patients seeking medical care for the condition that is being investigated. It is preferable to include as cases individuals who have been recently diagnosed with the illness rather than individuals who have had the disease for an extended period of time, in order to discriminate between exposure that occurred before disease onset and exposure that occurred after. Other sources of cases include disease registries, drug surveillance programs, schools, and places of employment (Kelsey, Thompson, & Evans, 1986).

Controls must be "representative of the same base experience" as the cases (Miettinen, 1985). "[T]he control series is intended to provide an estimate of the exposure rate that would be expected to occur in the cases if there were no association between the study disease and exposure" (Schlesselman, 1982: 76). It is important to recognize that controls are theoretically continuously eligible to become cases. An individual who is initially selected as a control and who later develops the disease(s) under study, thereby becoming a case, should be counted as both a case and a control (Lubin & Gail, 1984). Controls are frequently selected from probability samples of the population from which the cases arose; from patients receiving medical care at the same facilities as the cases, but for conditions unrelated to the cases' diagnoses; or from neighbors, friends, siblings or coworkers of the cases (Kelsey, Thompson, & Evans, 1986). Dead controls may also be used in studies where the researcher wishes to compare individuals who died from one cause with individuals who died from other causes (Lasky & Stolley, 1994).

Case-control studies are valuable because they permit the evaluation of a range of exposures that may be related to the disease under investigation. They are generally less expensive to conduct than cohort studies, in part because fewer people are needed for the study. However, it may be difficult to determine individuals' exposure status (Rothman, 1986).

Cross-Sectional Studies. Unlike either cohort studies or case-control studies, exposure and disease status are measured at the same point in time in cross-sectional studies. This approach results in a serious limitation, in that it may be difficult to determine whether the exposure or the disease came first, since they are both measured at the same time. Additionally, because cross-sectional studies include prevalent cases of a disease, *i.e.*, new cases and already-existing cases, a higher proportion of cases will have had the disease for a longer period of time. This may be problematic if people who die quickly or recover quickly from the disease differ on important characteristics from those who have the disease over a long period of time. Too, individuals whose disease is in remission may be erroneously classified as nondiseased (Kelsey, Thompson, & Evans, 1986).

Ecological Studies. In the study designs previously discussed, the individual was the unit of observation. Ecological designs, however, utilize a group of people, such as census tract data, as the unit of observation (Rothman, 1986). An example of an ecological study would be an examination of oral cancer rates against the use of chewing tobacco in each state. Ecologic studies are often conducted to observe geographic differences in the rates of a specific disease or to observe the relationship between changes in the average exposure level and changes in the rates of a specific disease in a particular population. They are useful in generating etiologic hypotheses and for evaluating the effectiveness of a population intervention. Because data are available only at the group level, however, inferences from the ecological analysis to individuals within the groups or to individuals across groups may be seriously flawed (Morgenstern, 1982). This often results from an inability to assess and measure extraneous factors on an individual level that may be related to the disease and the exposure under examination.

Measures

Clearly, an evaluation of causation in the context of epidemiology requires an evaluation of the study designs used, including the strengths and weaknesses of design. The evaluation must also consider the measures resulting from the studies. Epidemiology utilizes measures of disease frequency, which focus on the occurrence of disease; measures of association, which estimate the strength of the statistical association between a factor under study and the disease under study; and measures of impact, which help to explain the extent to which a particular factor has contributed to the cause or prevention of a specific outcome in a population. These measures have been discussed extensively elsewhere and will be reviewed here only briefly.

The incidence rate refers to the number of new cases of a disease in a specific population, divided by the total of the time periods of observation of all the individuals in the same population. Incidence rates can be calculated from the data obtained from cohort studies. Unlike incidence, which refers to the number of new cases that develop during a specified time period, prevalence relates to the proportion of a specific population that has the disease at a specific point in time or during a specific period of time (Kleinbaum, Kupper, & Morgenstern, 1982).

The concept of risk refers to the probability of developing a disease during a specific time period (Kleinbaum, Kupper, & Morgenstern, 1982). Probability can be thought of as a continuum between one and zero, where a value close to zero indicates that an event is unlikely to occur and a value close to one indicates that the event is more likely to happen. Probabilities are often expressed as percentages. Risk is approximately equal to the incidence rate multiplied by time, over a short period of time (Rothman, 1986).

Measures of association are obtained by comparing one or more study groups to another study group that has been designated as the reference group. The risk ratio, used with cohort studies, involves a comparison of the risk of disease in the group exposed to the factor of interest with the risk of disease in the group that has been unexposed to the factor of interest. The odds ratio, used with case-control studies, is derived from a comparison of the odds of having been exposed given the presence of the disease with the odds of not having been exposed given the presence of disease. This ratio is then compared to the odds resulting from a comparison of the odds of having been exposed given the absence of disease with the odds of not having been exposed given the absence of disease (Kelsey, Thompson, & Evans, 1986).

Cross-sectional studies often rely on the prevalence ratio, which results from a comparison of the prevalence of the disease in the exposed population and the prevalence of the disease in the unexposed population (Kelsey, Thompson, & Evans, 1986). It is unlikely that the prevalence ratio will be encountered very often in the context of litigation due to our inability to draw causal inferences from cross-sectional studies.

Ratio measures of association can be used to calculate measures of impact (Kleinbaum, Kupper, & Morgenstern, 1982). The attributable fraction, also called the attributable risk percent for the exposed, is calculated by subtracting the rate of disease among the unexposed from the rate of disease among the exposed and then dividing that difference by the rate among the exposed (Kelsey, Thompson, & Evans, 1986). The formula for this calculation is (rate for exposed)−(rate for unexposed)/rate for exposed. The attributable risk fraction, also known as the etiologic fraction, attributable risk, and population attributable risk (Kleinbaum, Kupper, & Morgenstern, 1982), is calculated by subtracting the rate of disease among the unexposed from the rate of disease among the entire population. This is written as (rate for the entire population)−(rate for unexposed)/rate for entire population. That difference is then divided by the rate for the entire population. The fraction that results depends on the prevalance of the exposure in the population and the magnitude of the rate ratio (Kelsey, Thompson, & Evans, 1986). Some writers have used the terms attributable fraction and attributable risk synonymously (Black, Jacobson, Madeira, & See, 1997).

Consider, as an example, the use of the risk ratio to calculate the attributable fraction. Suppose that a retrospective cohort study has indicated that exposure to a particular substance carries twice the risk of developing a specific form of cancer, compared to nonexposure, i.e. the relative risk is 2. The attributable fraction would be calculated by subtracting the risk of disease among the unexposed, which would be 1 thereby indicating no increased risk, from the risk of disease among the unexposed, 2. That difference is then divided by the risk among the exposed, 2. This yields an attributable fraction of .5.

Interpreting the Results

Statistical Significance and Confidence Intervals

In testing an hypothesis of association between exposure and disease, the investigator often begins with the hypothesis that there is no association between the exposure and the disease (null hypothesis). If the data do not support the null hypothesis, it can be rejected. The investigator must specify the level of statistical significance (alpha) which will indicate that any association that is found is unlikely to have occurred by chance. Although this alpha level is usually set at .05, this is completely arbitrary (Rothman, 1986). It is important to remember that statistical significance does not equate to clinical significance; a result may be clinically significant even in the absence of statistical significance.

A "p-value" is a statistic used to test the null hypothesis. It refers to the probability that the data will depart from an absence of association, to an extent equal to or greater than that observed, by chance alone, assuming that the null hypothesis is true. A lower p-value indicates a higher degree of inconsistency between the null hypothesis of no association and the data. Stated more simply, the p-value is the probability that we will make a mistake and reject the hypothesis of no association when, in fact, it is true. We want the p-value to be very small. The smaller the p-value, the more certain we are that the null hypothesis is not true.

An alpha, or Type I, error occurs when the null hypothesis is erroneously rejected, *i.e.*, it is true and it is rejected as false. If the p-value was set to .05, an alpha error will occur approximately five percent of the time. Conversely, a beta, or Type II, error will occur if the null hypothesis is false and is not rejected.

Reliance on p-values has been criticized because p-values fail to provide information about the magnitude of an effect estimate or its variability (Rothman, 1986). Rothman has urged, instead, the use of interval estimation (confidence intervals). A point estimate, which is a single best estimate of the parameter, *e.g.*, an effect measure such as an odds ratio, is derived from the data. A confidence interval, equivalent to one minus the alpha level, is calculated around the point estimate. As an example, if the alpha level is .05, the confidence interval to be constructed is a 95% confidence interval.

Internal And External Validity

The scientific validity of a study and, consequently, its legal reliability (see below), depends on its internal validity and its external validity, or generalizability. Internal validity refers to the validity of the inferences drawn that relate to the study participants. External validity refers to the validity of the inferences drawn as they relate to groups other than the study population, such as all adults, or all children with a particular disease.

Internal Validity. Biases in research can impact on a study's internal validity by affecting the accuracy of measurement. The *Dictionary of Epidemiology* defines "bias" as "any trend in the collection, analysis, interpretation, publication or review of data that can lead to conclusions that are systematically different from the truth" (Last, 1988: 13–14). There are many types of bias, but they are often classified into three general categories: selection bias, information bias, and confounding.

Selection bias may result from flaws in the procedures utilized to select participants for the study. These flaws lead to a distortion in the estimate of effect (Kleinbaum, Kupper, & Morgenstern, 1982). Selection bias might exist, for example, in a study of birth defects that relies on volunteers because the women who refer themselves to the study may be doing so for reasons related to the outcome (birth defects) under study.

Information bias results in a distortion of the effect estimate due to misclassification of the research participants on one or more variables. The misclassification may result from measurement errors or recall bias. Misclassification is said to be nondifferential if the misclassification on one axis (exposure or disease) is independent of misclassification on the other axis. Differential misclassification occurs when misclassification on one axis is not independent of misclassification on the other axis (Rothman, 1986). Differential misclassification can result in an over- or underestimation of the effect measure (Copeland, Checkoway, McMichael, & Holbrook, 1977). Recall bias is one form of differential misclassification that may occur in a case-control study that relies on participants' memories of their exposure experiences. Memory may differ between the exposed cases and the nondiseased controls for a variety of reasons. As an example, particular exposures may become more significant to the cases in retrospect because of an attempt to identify a cause or reason for the illness.

Confounding can occur if the exposure of interest is closely linked to another variable and to the disease of interest. For instance, if one were to study the association between alcohol use and a specific form of cancer without collecting data on levels of smoking, the results would be confounded if that form of cancer were associated with tobacco usage.

To be a confounder, a factor (1) "must be associated with both the exposure under study and the disease under study ..." (2) must "be associated with the exposure among the source population for cases;" and (3) may not be "a step in the causal chain between exposure and disease ..." (Rothman, 1982). It is important to note that, just as with exposure and disease status, confounders are subject to misclassification. Under some circumstances, such misclassification can seriously distort the results (Greenland & Robins, 1985).

Confounding is often confused with effect modification. Unlike confounding, which refers to a bias in the estimate of effect that results from a lack of comparability between the groups in the study, effect modification refers to a heterogeneity of effect. As an example, if the effect of smoking on cervical cancer differs by ethnicity, we would say that ethnicity in this case is an effect modifier. Ethnicity could also, however, be a confounder if there is an association between ethnicity and smoking among the controls.

External Validity. Scientific generalization is "the process of moving from time- and place-specific observations to an abstract universal statement" (Rothman, 1982: 96). As an example, the applicability to women of

Key: C=covariate D=disease E=exposure

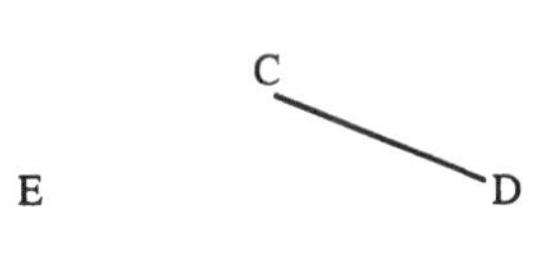

Figure 1-1 Factors C and D are associated in the study population, but not necessarily in the base population. This may result from selection procedures. In this situation, C is not a confounder.

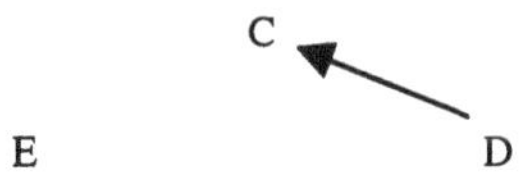

Figure 1-2 The C-D association results from the effect of D on C. C is not a confounder.

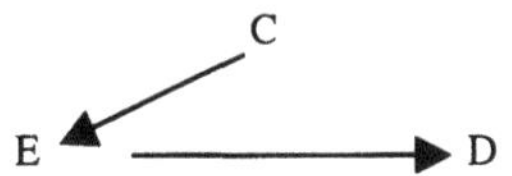

Figure 1-3 The effect of C on D in the base population is not independent of the exposure. C is not a confounder.

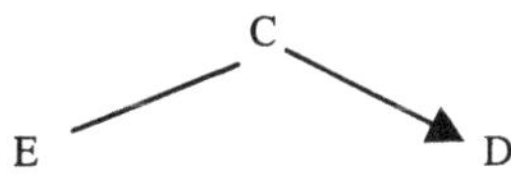

Figure 1-4 C is associated with exposure status in the base population and C is a risk factor for D in the unexposed base population. C is a confounder.

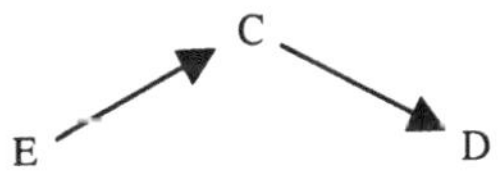

Figure 1-5 C is an intermediate (intervening) variable in the causal pathway between E and D. C is not a confounder.

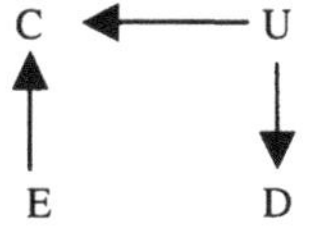

Figure 1-6 Both C and D are affected by the same unmeasured risk factor U and C is affected by E. C is not a confounder.

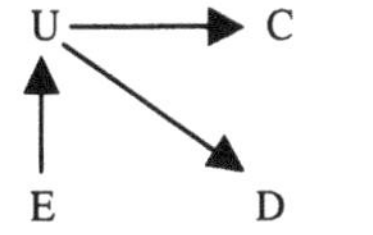

Figure 1-7 Both C and D are affected by another unmeasured risk factor, U. U is affected by E. C is not a confounder.

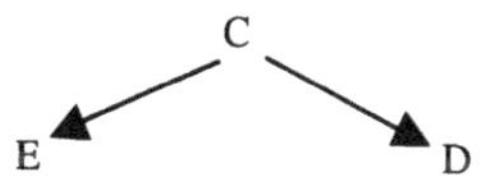

Figure 1-8 C is a risk factor for both E and D. C is a causal confounder.

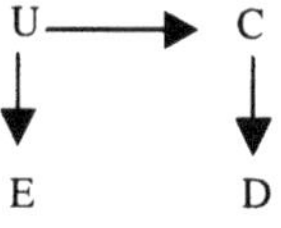

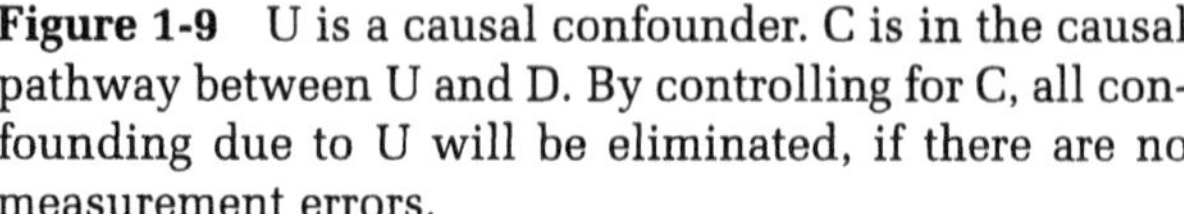
Figure 1-9 U is a causal confounder. C is in the causal pathway between U and D. By controlling for C, all confounding due to U will be eliminated, if there are no measurement errors.

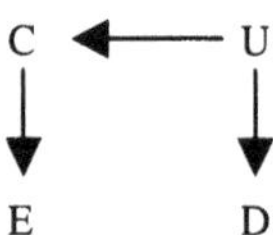

Figure 1-10 U is a causal confounder. C is in the causal pathway between U and E. By controlling for C, all confounding due to U will be eliminated, if there are no measurement errors.

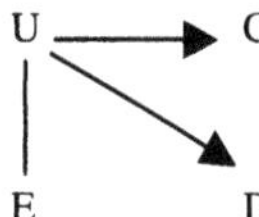

Figure 1-11 U is a causal or proxy confounder. C is associated with U. C is not in every causal pathway between U and E or between U and D. Controlling for C will will not eliminate confounding by U.

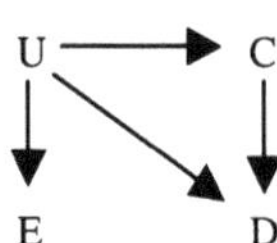

Figure 1-12 U is a causal or proxy confounder. C is associated with U, but is not in every causal pathway between U and E or U and D. Controlling for C could result in either increased or decreased bias because the direct and indirect effects of U on D could be in opposite directions.

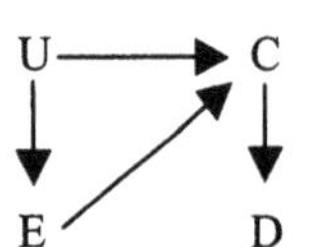

Figure 1-13 C is a proxy for the unmeasured confounder U. Confounding due to U will be eliminated by controlling for U if C is an intermediate variable and U is a confounder. If we have no information on U, C is a proxy confounder in addition to being an intermediate variable. Our estimate of E will be biased whether or not we control for C.

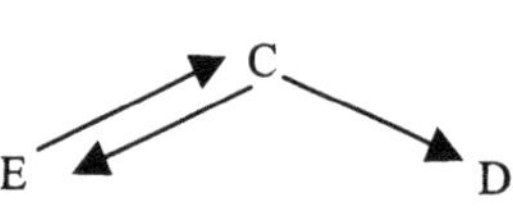

Figure 1-14 C is a time-dependent variable and is both a confounder and an intermediate between E and D. Conventional statistical methods will not produce an unbiased estimate of the effect of E.

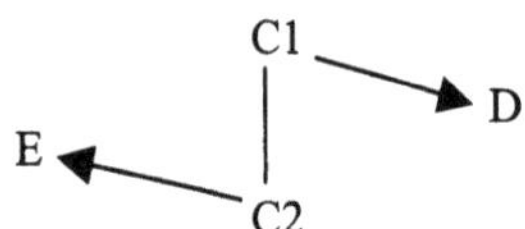

Figure 1-15 C2, a confounder, is a proxy for C1, a confounder. C1 and C2 are redundant confounders. Analysis must control for either C1 or C2 to eliminate bias.

the results of clinical trials relating to heart disease and HIV in men, for example, has been called into question (Mastroianni et al., 1994).

Strategies to Increase Validity. A variety of options are available to control for potential bias. One such option, randomization, was discussed previously in the context of clinical trials and that discussion will not be repeated here. Other options include restriction, matching, stratification, and mathematical modeling.

Restriction is used in the design phase of a study to limit inclusion in the study to individuals who meet certain predetermined eligibility criteria. In this way, individuals who possess certain characteristics may be excluded, thereby reducing the potential of bias due to the presence of those extraneous factors (Gray-Donald & Kramer, 1988). As an example, investigators studying the etiology of HIV-associated dementia would exclude from the study individuals who possess other risk factors for dementia or dementia-like conditions, such as current alcohol or drug abuse or certain forms of mental illness. As a technique, restriction is relatively inexpensive and facilitates the analysis and interpretation of the results. However, generalization to populations other than those included in the study may not be valid. Additionally, restriction may not control completely for confounding (Kleinbaum, Kupper, & Morgenstern, 1982). Restriction may also raise ethical issues due to the routine exclusion of specific groups, such as women, from participation in research studies (Mastroianni et al., 1994).

"Matching" refers to the selection of a comparison group that is comparable to the study group with respect to certain prescribed characteristics that could potentially confound or bias the results, such as age or sex. Matching reduces or eliminates the potential variability between the study and comparison groups with respect to the "matched" variables. Consequently, the factors selected as the basis for matching should not be factors of interest that the investigator wishes to examine further (Kelsey, Thompson, & Evans, 1986). For example, an investigator may be interested in factors that affect the quality of care received by HIV-infected patients. If the investigator matches on insurance status, he or she will be unable to investigate the effect that insurance has on whether or not the patient receives a particular treatment or undergoes a particular procedure. Although matching can be used with cohort, case-control, and cross-sectional designs (Kelsey, Thompson, & Evans, 1986), it is most often used in case-control designs. Matching is often not feasible in cohort and cross-sectional studies due to lack of information on the potentially confounding factors (Kelsey, Thompson, & Evans, 1986).

The variables on which cases and controls are to be matched must be thought to be related to both disease and exposure, *i.e.*, they must be confounders that must be controlled for in the design or the analysis. Additionally, matching should not be unduly costly (Kelsey, Thompson, & Evans, 1986). Cost may become a particular issue if matched controls are difficult to locate due to the closeness of the matching that is required (Smith, 1983). Matching brings several advantages, including the ability to more adequately control for confounding variables, the ability to obtain time comparability between cases and controls for exposures that may vary over time, and a gain in statistical power (Wacholder, Silverman, McLaughlin, & Mandel, 1992).

Stratification means dividing the analysis into two or more groups, such as males and females, to permit separate analysis of each. Stratified analysis can be used in cohort, case-control, and cross-sectional studies. This technique requires that levels of the confounding variables be defined and the exposure-disease association estimated within each level. Stratification is appropriate if (1) there is a sufficient number of individuals in each level or strata; (2) the choice of control variables is appropriate; and (3) the definition of levels for each of the confounding variables is appropriate (Kleinbaum, Kupper, & Morgenstern, 1982). As an example, smoking could be a confounding variable in a study examining the relationship between asbestos exposure and certain lung diseases. Consequently, the investigator would want to stratify the analysis by level of smoking, *e.g.*, lifetime levels of smoking.

Mathematical modeling involves the use of analyses that relate exposure, outcome, and extraneous variables. These analyses are said to be multivariate because they include multiple factors in the model. In cohort studies, the illness or disease outcome or status is often the dependent variable, while either the exposure status or disease outcome may be the dependent variable in case-control studies. Mathematical modeling has many advantages, including ease of use with small numbers; the ability to predict individual risk; and the ability to use this technique with continuous variables and with multiple exposure variables.

Mathematical modeling may also entail several disadvantages. All models require that certain assumptions be made about the data prior to the application of the model. The selection of the model requires an evaluation of these underlying assumptions. If the assumptions do not hold, another modeling technique must be used. Mathematical modeling may also present difficulty in the interpretation of the results (Kleinbaum, Kupper, & Morgenstern, 1982).

Sampling and Sample Size. Many epidemiologic studies utilize data from already existing data sources, often collected on a routine basis, such as data from various disease registries or hospital discharge records. This type of data is known as secondary data. It is crucial that these data sources be as accurate as possible, e.g., that minimal misclassification exists.

Most epidemiologic studies, however, rely on primary data collected from the original source, such as individuals with a particular disease or exposure of interest. This often requires the identification of the individuals who will participate, the development of questionnaires, and/or the conduct of interviews. In most cases, a sample of such individuals must be used, rather than an entire population. Reliance on a sample of the population helps to reduce study costs and, to some extent, may increase the accuracy of the measurements since more time can then be focused on fewer people.

When assembling a sample population, the sampling unit will depend on the particular study. Most often, the unit will be an individual or a household, although it may also be a neighborhood center, school, or other entity. A listing of the sampling units constitutes the sampling frame.

Sampling Techniques. The manner by which sampling is conducted is critical. Appropriate sampling techniques will reduce random error, which can reduce the precision of the epidemiologic measurements that are obtained (Rothman, 1986). Probability sampling is one method of reducing random error. Probability sampling refers to sampling by which "each sampling unit has a known, nonzero probability of being included in the sample" (Kelsey, Thompson, & Evans, 1986). There are various techniques available for probability sampling, including simple random sampling, systematic sampling, stratified sampling, cluster sampling, and multistage sampling. Other sampling techniques, such as snowball sampling, are available for use in situations where probability sampling is not possible.

With simple random sampling, each sampling unit in the eligible population has an equal chance of being included in the study. In order to conduct random sampling, the investigator must know the complete sampling frame, *i.e.*, everyone who is potentially eligible to be in the study. Sampling can occur with or without replacement. With replacement sampling, selected sampling units, such as the individuals selected for participation in the study, are returned to the pool from which the sample is being taken. Most epidemiologic studies rely on sampling without replacement, which yields more precise estimates.

With systematic sampling, sampling units are selected from regularly spaced positions from the sampling frame, such as every tenth patient admitted to an inpatient facility. Systematic sampling is relatively easy to implement and does not require *a priori* knowledge of the sampling frame.

Stratified sampling requires that the population be divided into predetermined strata. Within each strata, the sampling units share particular characteristics, such as sex. The study participants are then selected by taking a random sample from within each strata. Stratified sampling is particularly useful to ensure that all subgroups of interest are represented in the study population. Stratified sampling may also yield more precise estimates of the population parameters since the overall variance is based on the within-stratum variances (Kelsey, Thompson, & Evans, 1986).

Disproportionate stratification refers to the disproportionate sampling of strata, such as specific neighborhoods, which contain high concentrations of the population of interest. For instance, if an investigator were interested in studying the effect of culture on nutritional intake, he or she might over-sample certain groups to ensure that there is a sufficient number of such individuals in the study to be able to analyze the data. This technique results in unequal selection probabilities for members of the different strata, thereby requiring weighting adjustments in the analysis of the data (Kalton, 1993).

Cluster sampling involves the selection of clusters from the population. Observations are then made on each individual within a cluster. As an example, one may wish to identify certain neighborhoods (clusters) and then sample all households within those selected neighborhoods.

Multistage sampling is similar to cluster sampling in that it first requires the identification of primary sampling units, such as neighborhoods. Unlike cluster sampling, however, multistage sampling utilizes a sample of secondary units with each primary unit, rather than a sample of all of the secondary units. As an example, multistage sampling would require the sampling of households within the selected neighborhoods, such as every tenth house in the selected neighborhoods, rather than reliance on all of the households within each selected neighborhood (Kalton, 1993).

Location sampling refers to the selection of participants through recruitment at locations and times when numbers of the target population are expected to be high. Locations could include, for instance, bars, grocery stores, bookstores, or churches, depending upon the population to be sampled. This type of sample is generally considered a convenience sample.

Time/space sampling is conducted at specified locations and times when population flows are expected to occur, such as voting booths on election day. A sampling frame consisting of time/location combinations is constructed and a sample of individuals is then selected from these selected sampling units. This method of sampling produces a probability sample of visits, rather than individuals (Kalton, 1993).

Snowball sampling is based on the premise that members of a particular, *i.e.*, rare, population know each other. Individuals are identified within the targeted rare population. These individuals are asked to identify other individuals within the same target group. This technique can be used to generate the sample ("snowball sampling") or, alternatively, to construct a sampling frame for the rare population, from which the sample is then selected. For example, if an investigator wished to examine the prevalence of needle sharing behaviors among injection drug users, snowball sampling would permit the investigator to identify eligible participants by relying on previously identified eligible participants. This could be more efficient than attempting to identify individuals through hospitals or clinics. Snowball sampling is a nonprobability sampling procedure. Snowballing for frame construction does not suffer from this weakness, but carries the possibility that socially isolated members of the rare population will be missing from the frame (Kalton, 1993).

Sample Size and Power. The epidemiologist is concerned not only with the appropriateness of the sampling method used, but also with the size of the sample. Increasing the size of the sample may reduce random error and increase precision.

Power calculations are often utilized to determine the requisite sample size. "Power" has been defined as "the probability of detecting (as 'statistically significant') a postulated level of effect" (Rothman, 1986). In order to calculate power, one must specify the (statistical) significance, or alpha level; the magnitude of the effect that one wishes to detect, such as an odds ratio of 2; the sample size of the exposed group in a cohort or cross-sectional study or the sample size of the diseased group in a case-control study; and the ratio of the size of the comparison group to the size of the exposed group in a cohort or cross-sectional study or to the size of the diseased group in a case-control study (Kelsey, Thompson, & Evans, 1986).

An alternative approach to assessing the adequacy of the sample size is to calculate the requisite size using one of the accepted sample size formulas. These formulas require that the investigator specify the level of statistical significance (alpha error); the chance of missing a real effect (beta error); the magnitude of the effect to be detected; the prevalence of the exposure in the nondiseased or the disease rate among the unexposed; and either the ratio of the exposed to the unexposed or the ratio of the cases to the controls (Rothman, 1986). Reliance on such formulas has been criticized because they create the illusion of a boundary between an adequate and inadequate sample size when, in fact, the variables specified to determine the sample size are often set either arbitrarily or by relying on estimates (Rothman, 1986).

Meta-analysis offers "[a] quantitative method of combining the results of independent studies (usually drawn from the published literature) and synthesizing summaries and conclusions which may be used to evaluate therapeutic effectiveness, plan new studies, etc ... " (Olkin, 1995: 133). Meta-analysis refers to not only the statistical combination of these studies, but also to "the whole process of selection, critical appraisal, analysis, and interpretation..." (Liberati, 1995: 81). Because meta-analysis permits the aggregation of studies' results, it may be useful in detecting effects that have been somewhat difficult to observe due to the small sample size of individual studies. Various approaches to meta-analysis have been subject to criticism, including the use of quality scores for the aggregation of studies of both good and poor quality and over-reliance on p-values (Olkin, 1995). Additionally, because "[t]he validity of a meta-analysis depends on complete sampling of all the studies performed on a particular topic" (Felson, 1992: 886), the results of meta-analysis may be biased due to various forms of sampling bias, selection bias, or misclassification (Felson, 1992). The establishment of more or less rigid inclusion and exclusion criteria can impact heavily on the results of a meta-analysis.

Causation in Law

It is important to recognize that

> [a]ll scientific work is incomplete—whether it be observational or experimental. All scientific work is liable to be upset or modified by advancing knowledge. This does not confer upon us freedom to ignore the knowledge we already have, or to postpone action that it appears to demand at a given time (Hill, 1965).

As demonstrated below, in the legal context, prompt action may be necessary even when we lack perfect, or even extensive, knowledge.

Sources of Law

Law is frequently classified into two domains, the public and the private. Public law encompasses law that is concerned with government or its relations with individuals and businesses. Public law is concerned with the definition, regulation, and enforcement of rights where an entity of the government is a party to the action. Public law derives from constitutions, statutes, and regulations and rules that have been promulgated by an administrative entity, such as a federal agency. For instance, the regulations of the Food and Drug Administration with respect to informed consent would be classified as public law.

Private law refers to law that regulates the relations between and among individuals and individuals and businesses. This includes actions relating to contracts, to property matters, and to torts. The primary sources of private law include statutes and judicial decisions.

Law is also classified into criminal and civil law. Criminal law deals with crimes. Even though a crime may have been committed against a person, for instance, when a person is robbed, the crime is said to have been against the state and it is the state (or federal government, depending upon the nature of the crime and the basis of the charge) that has the right to prosecute the accused individual or entity. Civil law is that law that refers to non-criminal public and private law.

The sources of law can be thought of as being in an inverted pyramidal shape. At the very base of this inverted pyramid is the constitution. Everything above the constitution must be consistent with the principles enunciated in the constitution. Above the constitution are the statutes. As you move up the inverted pyramid, you find the regulations and the precepts that have been derived from cases heard by the court. At each level, the decisions and principles must be consistent with those of the previous levels. Although it would seem that the system is relatively unstable because the constitution, which forms the basis for everything else, is at the point of the pyramid, it is actually quite stable because everything else must remain in balance with the constitution.

The Constitution

The federal constitution has been called the "supreme law of the land." The Constitution actually represents a grant of power from the states to the federal government because all powers not specifically delegated to the federal government are, pursuant to the terms of the Constitution, reserved to the states.

The Constitution allocates power among three branches of government. The legislative branch is charged with the responsibility and delegated the authority to make laws (statutes). The executive branch of the government is responsible for the enforcement of the laws, while the judicial branch is responsible for the interpretation of those laws.

There are 26 amendments to the main body of the Constitution. The first 10 of these amendments are known as the Bill of Rights. These encompass many of the rights with which people may be most familiar, such as freedom of speech and freedom of religion. It is important to remember, though, that these rights as delineated are in the federal constitution and as such apply to the federal, not state, government. The Fourteenth Amendment, however, provides specifically that no state may deprive any person of life, liberty, or property without the due process of law. The Amendment also provides that no state may deny equal protection to any person within its jurisdiction. Most of the rights that are enumerated in the Bill of Rights have been found by the Supreme Court to constitute due process, so that ultimately, these rights also apply to the states as well as to the federal government.

Each of the 50 states also has its own constitution. The state constitutions cannot grant to persons fewer rights than are guaranteed to them by the federal constitution. However, they may grant more rights than are provided for by the federal constitution.

Statutes

Statutes at the federal level are promulgated by Congress, consisting of the House of Representatives and the Senate. At the state level, the state legislatures, also consisting of two houses, are responsible for the promulgation of statutes. For example, Congress passed the laws which give the Food and Drug Administration and the Department of Health and Human Services their authority to make regulations. Judges are responsible for the interpretation of the statutes where there is a lack of clarity or where there is conflict between various statutory provisions.

Administrative Law

Administrative law is that law that is made by the agencies which comprise a part of the executive branch of government. Administrative law encompasses regulations, rules, guidelines, and policy memoranda. Examples of administrative agencies relevant to the health research context include the Food and Drug Administration, the National Institutes of Health, and the Department of Health and Human Services. Stated simplistically, an agency's regulations are developed and promulgated through a notice and comment procedures, whereby the proposed regulation is published in the *Federal Register*, which is available to the public for review. Following a mandated time period during which the promulgating agency may receive comments on its proposed regulation, the comment period will cease. After reviewing the comments and incorporating those that the agency deems appropriate, the agency will issue its final regulation. A similar process is followed on the state level. Administrative law is discussed more fully in chapter 4, which addresses the use of epidemiology in the formulation of agency-promulgated regulations.

Court Decisions

As indicated, judicial decisions must be consistent with statutes and the Constitution. The courts adhere to the doctrine known as *stare decisis*, meaning that they must look to past cases with similar facts and legal issues to resolve the cases that appear before them. In general, they are bound by decisions of all higher courts within the same jurisdiction. This will become clearer following a discussion of the structure of the legal system. For instance, all federal and state courts are bound by the decisions of the Supreme Court of the United States. All federal district courts are bound by the decisions of the federal Court of Appeal for the circuit in which the federal district court sits, but they are not bound by the decisions of a Court of Appeal for a different circuit. For instance, California sits in the Ninth Circuit. The federal district court for the southern district of California is bound by the decisions of the Ninth Circuit Court of Appeal, but is not bound by the decisions of the Fifth Circuit, which covers the geographic area encompassing such states as Texas and Louisiana.

Judicial decisions also follow the doctrine of *res judicata*. This means that once a case had been decided and all of the channels for appeal have been utilized, the party bringing the case may not bring it again.

The Structure of the Legal System

The state and federal court systems can be thought of as pyramids. At the very base of the pyramid are the lowest courts. At the mid-level of the pyramid sit the courts of first appeal and, at the pinnacle of the pyramid, sits the supreme court of the state or of the federal court system. Different states, however, name these various levels differently. For instance, the supreme court in California is known as the Supreme Court, but in Massachusetts it is known as the Massachusetts Supreme Judicial Court, and in New York it is called the Court of Appeals.

In the state court system, the lowest level courts are often divided into those that have limited jurisdiction and those that have general jurisdiction. Those with limited jurisdiction often hear cases involving less serious offenses and civil lawsuits that do not involve large sums of money. Courts of general jurisdiction may hear cases involving monetary sums over a specified amount or more serious matters. Courts of general jurisdiction are often divided into special courts due to the volume of cases and the need for specialized expertise. Examples of such specialized courts include juvenile court, family court, and family court.

The mid-level courts, or appellate courts, have the power to hear appeals from the decisions of the lower courts. This is known as appellate jurisdiction, as contrasted with original jurisdiction, which is the power to hear a case at its inception. The appellate courts may have original jurisdiction with respect to a limited range of cases. The state supreme court may hear appeals from the appellate courts.

The lowest tier on this pyramid in the federal system consists of the federal district courts. These courts hear cases involving crimes that arise under federal statutes, such as making false statements on a federal application. They have jurisdiction over cases in which the citizen of one state is suing a citizen of another state (diversity of citizenship case) if the amount in dispute is greater than a specified minimum. (State courts may also hear cases in which a citizen of one state is suing a citizen of another state. This is known as concurrent jurisdiction. Not infrequently, the party who did not file the original lawsuit may ask to have the case removed to federal court.) The federal district courts may also hear cases arising under the federal constitution and cases arising under federal statutes.

Appeals from the decisions of the district courts are made to the federal Court of Appeal having jurisdiction over the circuit in which the district court sits. There are 13 Courts of Appeal. Twelve of these are for named circuits, one is for cases arising in the District of Columbia, and the 13th is for the Federal Circuit, which has jurisdiction over claims that are exclusively within the domain of federal law, patent and trademark law.

The Supreme Court hears appeals from the Courts of Appeal. However, in most situations, there is no automatic right to appeal to the Supreme Court. Rather, the Supreme Court chooses the cases that it will hear. Request to have an appeal heard is made through a *writ of certiorari*, which is a petition to file an appeal.

Apart from the judicial system, some agencies may have the power to resolve cases administratively. For instance, the Office of Research Integrity has the authority to investigate and adjudicate allegations of scientific misconduct. Appeals proceed to the Department Appeals Board and, from there, to court if necessary.

Civil Procedure

A lawsuit is commenced through the filing of a complaint by a party to the lawsuit. The complaint must, in general, state the nature of the claim, the facts to support the claim, and the amount in controversy. The defendant will be served with a copy of the complaint, together with a summons. The summons indicates that the defendant must respond to the complaint in some fashion within a specified period of time or the plaintiff will win the lawsuit by default.

The defendant will then answer the complaint, and will admit, deny, or plead ignorance to each allegation of the complaint. The defendant may also file a countersuit against the plaintiff or against a third party. The defendant may also ask that the court dismiss the plaintiff's action, claiming that the court has no jurisdiction to entertain the case or that the plaintiff failed to state a cause of action.

Following the initiation of the lawsuit and the answer by the defendant, there will be a period of discovery, during which each party to the action will have the opportunity to gather additional facts to support its case, to identify expert witnesses that the other side may call, and to identify weaknesses in the opposing party's case. Discovery may include depositions, written interrogatories, the production of documents, a request for a mental or physical examination, and a request for admissions. Those that are most relevant to the health research context are depositions, written interrogatories, a request for the production of documents, and a request for admissions.

The trial itself consists of numerous stages:

1. the opening statement of the plaintiff,
2. the opening statement of the defendant,
3. the presentation of direct evidence by the plaintiff, with cross-examination of each witness by the defendant, re-direct by the plaintiff, and re-cross by the defendant,
4. the presentation of direct evidence by the defendant, with cross-examination by the plaintiff, re-direct by the defendant, and re-cross by the plaintiff,
5. presentation of rebuttal evidence by the plaintiff,
6. presentation of rebuttal evidence by the defendant,
7. plaintiff's argument to the jury,
8. defendant's argument to the jury,
9. plaintiff's closing argument to the jury,
10. instructions from the judge to the jury, and
11. jury deliberation and verdict.

Proving Causation in Negligence Actions

The sufficiency of data sufficient to establish a causal association between an action, injury, or exposure and a consequence will vary depending upon the relevant legal context. For instance, what is deemed to be sufficient in the context of formulating regulations to control the use of a pesticide that is believed to be carcinogenic may be quite different than the nature and/or quantity of the information or data that the jury finds necessary to establish that a plaintiff's physical illness resulted from exposure to the defendant's product, such as a pharmaceutical. This discussion focuses on the use of epidemiological data to establish causation in the context of litigation. The use of epidemiological data to establish causation in the context of formulating agency regulations is addressed in chapter 4.

In order to establish causation in the context of a civil lawsuit, the plaintiff must establish that the defendant owed a duty to the plaintiff, that that duty was breached, that the breach of that duty resulted in harm to the plaintiff (cause in fact), that there was a nexus between the defendant's conduct and the plaintiff's injury (proximate cause), and that the plaintiff is seeking damages. Legal jurisdictions differ with respect to whether a duty is owed to plaintiffs who may not be foreseeable. In some jurisdictions, if a duty is owed to anybody, it is owed to everybody. In other jurisdictions, duty is not owed to those who are unforeseeable. Even where, however, the plaintiff may be foreseeable, there may not be a duty owed depending upon the closeness of the connection between the defendant's conduct and the harm that the plaintiff suffered.

Establishing a Breach of a Duty

In order to establish that the defendant breached a duty of care, the plaintiff must provide proof of what actually happened and must demonstrate that the defendant acted unreasonably under the circumstances. In determining whether the conduct is unreasonable, the court may look at the balance between the risks and the benefits of the defendant's conduct. The risk refers to the severity of the harm that might occur as a result of the defendant's conduct and the probability that that harm will occur. In evaluating the benefit, the court may consider such things as the existence and availability of safer alternative methods, the costs of these alternative methods, and the social value attached to the defendant's conduct. The defendant's conduct will be considered unreasonable if a reasonable person in the defendant's position would have perceived in advance that the risks of the conduct outweighed its benefits.

Establishing Cause in Fact

There are several different rules, or standards, to determine whether the conduct of the defendant was responsible for the injuries suffered by the plaintiff. Different jurisdictions follow different rules. As a result, recovery by a plaintiff against a defendant may depend on the jurisdiction in which the harm occurred and the lawsuit is filed, even given the same facts.

The "but for" rule basically says that the plaintiff would not have been injured, but for the conduct of the defendant. This standard is essentially the legal equivalent of the deterministic model of causation in epidemiology.

A second rule of causation is that of concurrent liability. For instance, if the plaintiff is injured through the actions of a defendant, together with the actions of a third party, and the plaintiff would not have been injured but for the concurrence of the actions of the defendant and the third party, then both the defendant and the third party will be said to be the actual cause of the injury.

The third rule of causation is that of the substantial factor. Assume that a plaintiff suffers an injury due to the conduct of the defendant and a third party. If the defendant's conduct was a substantial factor in bringing about the injury, he or she will be found to have caused the plaintiff's injury. This is the legal equivalent of modified determinism in epidemiology, as described by Rothman (1986), whereby multiple factors may combine to bring about a result. If this type of fact situation were to occur in a jurisdiction that utilized a "but for" rule, both the defendant and the third party would be found not to be liable because in neither instance could it be demonstrated that but for the conduct, the plaintiff would not have been harmed.

Some jurisdictions have accepted the principle of alternative liability. In such cases, there may be several defendants, each of whom committed the same conduct. However, it is impossible to determine whether it was the conduct of one or the other that resulted in harm to the plaintiff. This rule has led to the development of what is known as market share liability, which has been used in the context of a number of health-related cases.

For instance, consider the use of diethylstilbesterol (DES) and the resulting harm. Various manufacturers may have made the product, but it may be impossible at the time that the lawsuit is filed for a plaintiff to know which manufacturer made the DES that she had ingested, and the DES manufactured by the various companies was essentially indistinguishable between those companies. New York has adopted the view that liability is related to the defendant's national market share of the product. The defendant will be found not to be liable only if it can show that it did not produce the product for the use that injured the plaintiff (*Hymowitz v. Eli Lilly*, 1989). In contrast, California finds that a defendant will be liable for the percentage of the plaintiff's injuries that is equal to the market share held by the company (*Sindell v. Abbott Laboratories*, 1980).

Some jurisdictions rely on a fifth rule, known as loss of a chance. In such instances, the plaintiff must demonstrate that he or she has lost something that he or she more likely than not would have retained or acquired, but for the conduct of the defendant. For instance, some courts have allowed a plaintiff to recover against the defendant where the plaintiff was physically hurt and now fears further harm, such as the development of cancer (*Mauro v. Raymark Industries, Inc.*, 1989).

Establishing Proximate Cause

The term "proximate cause" is actually a misnomer, because it refers not to cause, but to a policy decision: to which consequences of his or her conduct can the defendant's liability be extended? This determination is often made with reference to the foreseeability of the plaintiff, the foreseeability of the manner in which the breach of the duty occurred, and the foreseeability of the result. It is important to remember that, if there was no duty owed to the plaintiff, the issue of foreseeability is never reached. Other factors may also be considered, such as whether there were intervening acts that either extended the results of the defendant's conduct or combined with the defendant's conduct to produce the harm suffered by the plaintiff.

Assessing the Reliability of the Scientific Evidence

Expert testimony in the area of epidemiology is permitted if the testimony is scientific or technical or involves other specialized knowledge and if the testimony will "assist the trier of fact to understand the evidence or to determine a fact in issue" (Federal Rule of Evidence 702). The epidemiologist can provide his or her testimony in the form of an opinion or an inference. In order to testify, however, the epidemiologist must be qualified as an expert "by knowledge, skill, experience, training, or education ..." (Federal Rule of Evidence 702).

In order to be considered scientific knowledge for the purpose of testifying, the evidence to be proffered must be both relevant and reliable.

Evidence is relevant if it has a "tendency to make the existence of any fact that is of consequence to the determination of the action more probable or less probable than it would be without the evidence" (Federal Rule of Evidence 401). Evidence that is not relevant will not be admitted. Additionally, evidence that is relevant will be excluded "if its probative value is substantially outweighed by the danger of unfair prejudice, confusion of the issues, or misleading the jury, or by considerations of undue delay, waste of time, or needless presentation of cumulative evidence" (Federal Rule of Evidence 403). For instance, in *In re Bendectin Litigation* (1988), in which damages were sought against the manufacturers of the drug Bendectin for injuries alleged to have been caused to the children of mothers who ingested Bendectin during their pregnancies, the court refused to admit evidence relating to the drug thalidomide, manufactured by the same company, because of the danger of unfair prejudice.

In order to be considered reliable, the scientific testimony "must be supported by appropriate validation" (*Daubert*, 1993: 113 S.Ct. at 2787). An assessment of the (legal) validity of the scientific testimony requires that the court (1) determine whether the theory or technique at issue could have been tested; (2) determine whether the methodology has been subjected to peer review; (3) consider the potential rate of error associated with a particular method or technique; and (4) assess the extent to which the proffered methodology has been accepted within the identified relevant scientific community (*Daubert*, 1993).

Various mechanisms can be utilized to determine whether the proffered epidemiological evidence is (legally) reliable. These include (1) a full pre-trial hearing to assess the reliability of the evidence; (2) an examination of the pleadings, depositions, answers to interrogatories, admissions on file, and affidavits in the context of a motion for summary judgment; and (3) direct- and cross-examination in the context of the trial itself. The court may also appoint its own expert to examine the evidence to be offered.

A full pre-trial hearing on the admissibility of the scientific evidence is known as an in limine *Daubert*-type hearing. This type of hearing occurs before the actual trial and focuses on the qualifications of the experts who will testify, the nature of the areas of expertise, their examination of already-existing studies, and the validity of current scientific studies. If the judge finds in the context of this in limine hearing that the scientific evidence is unreliable, and therefore inadmissible, neither the plaintiffs nor the defendants would be able to use that evidence at trial.

Under the Federal Rules of Civil Procedure, interrogatories may be directed only to a party to the action. Interrogatories are written questions addressed to the opposing party, to which they will respond in writing. Under the current Rules of Civil Procedure, the expert witness-epidemiologist who has been retained to provide expert testimony in a case must prepare and sign a written report, which must include a statement of the expert's opinions, including the basis for those opinions; the data or information that the expert used in forming his or her opinions; any exhibits that the expert will be using; a statement of the expert's qualifications; and a list of any other cases during the previous four years in which the epidemiologist-expert testified as an expert either at trial or by deposition (Federal Rule of Civil Procedure 26). It is believed that the production of this report will reduce the need for interrogatories and the length of depositions. The following questions are part of a set of interrogatories that were developed for use in a lawsuit relating to injuries alleged to have arisen from the use of silicone breast implants.

> These interrogatories seek to obtain information regarding epidemiologic human studies which you are: a) conducting; b) assisting; c) financing (directly or indirectly); d) familiar with; e) or to your knowledge are currently underway...which are designed in whole or in part to explore the causal relationship or statistical association between exposure to silicone or saline breast implants...and subsequent local or systemic disease or complaints, including but not limited [to] encapsulation, chronic inflammation, granuloma, rupture, connective tissue disease, auto-immune disease or any other pathology.
>
> As to each study, please identify and state the following:...
>
> 4. The number of subjects and controls, with a description of the study population(s) (exposed and unexposed in cohort study or cases and controls in case-control study). Include all exclusion criteria.... .
>
> 5. The methodology or design of the study whether: retrospective or prospective; cohort or case control; and the anticipated biases or confounding factors considered, and the manner or method employed to avoid or control them, including how exposure classification will be made (e.g., subject, interview, physical exam, medical records), and methods utilized to minimize exposure misclassification.
>
> 6. The principal interest, focus or hypotheses under consideration ... (*In re Silicone Gel Breast Implant Products Liability Litigation*, 1994).

A deposition is a record of testimony that is taken outside of the courtroom (*Windsor Shirt Company v. New Jersey National Bank*, 1992). Unlike interrogatories, a deposition can be directed to an expert witness, such as an epidemiologist. Questions during a deposition usually focus on four areas: the expert's qualifications as an expert, the procedures that the expert relied upon to form and render a professional opinion, the expert's opinion itself, and the process by which the expert arrived at his or her opinion (Matson, 1994). The following excerpt from the deposition of an expert witness in a lawsuit to recover damages alleged to have been caused from the use of silicone breast implants provides an example of the kinds of questions that would be asked of an epidemiologist in attempting to establish the scientific validity and, accordingly, the legal reliability, of the proffered evidence.

> Q.: Have you concluded for yourself or have you formed an opinion for yourself as to whether or not based upon the data that you've seen there is a statistically significant association between disease or disease process and the use of breast implants?
>
> A.: For some diseases, yes.
>
> Q.: And which diseases are those, please, sir?
>
> A.: Breast cancer being one, scleroderma being another one. In a general sense, connective tissue disease. I would also have to say there are certain local complications that I have...come to an opinion on.
>
> Q.: What are those, please, sir?

A.: Capsule formation, capsule hardening, contracture, infection associated with surgery, local complications ... .

Q.: You told me that as far as you were concerned that there are some diseases that bore a statistically significant association with the use of breast implants.

A.: No, I don't think that was the sense of my answer. You asked me whether I had formed an opinion about various diseases, and I said yes. For example, with breast cancer, my opinion is that there is no cause and effect related to breast implants. For scleroderma, it is my opinion that there is no cause and effect or no statistical association between scleroderma and breast implants. That's different than the question that you just posed to me.

Q.: All right. I want to talk with you just briefly about these epidemiological studies that you listed for me. And I've got approximately twelve. Describe for me the methodology in these various studies. (Cook, 1994)

Either the plaintiff or the defendant can bring a motion for summary judgment. This is a pretrial motion. The court can grant the motion only if "there is no genuine issue as to any material fact" (Federal Rule of Civil Procedure 56). If the party bringing the motion has the burden of proof on an issue raised by the motion, the court may grant the motion only if the evidence supporting the moving party's position is sufficient to permit a rational factfinder to find for the moving party under the appropriate standard. (The burden of proof refers to the facts that the plaintiff must prove in order to establish a *prima facie* case, such as duty, breach of duty, and harm, among others.) If the moving party does not have the burden of proof, the court may grant summary judgment only if the opposing party that does have the burden of proof does not produce evidence sufficient to permit a jury to find in its favor. Although the epidemiologist-expert witness is unlikely to testify in conjunction with a motion for summary judgment, he or she may be involved by helping a party to the action to identify relevant issues.

Direct examination of the epidemiologist-expert at trial is conducted by the attorney for the party for whom the epidemiologist is testifying. The direct examination generally addresses the expert's credentials, the methodology that the expert used to evaluate the relevant epidemiological research, and the methods used to conduct the studies in which the expert was involved. Direct testimony of the epidemiologist-expert provides an opportunity to educate the jury regarding the subject matter at hand, the relevant methodology, and the peer review process (Karns, 1994). Voke (1994: 47–48) has suggested that direct examination focus on twelve key objectives:

1. the establishment of the expert as someone who is highly qualified with regard to the subject matter;
2. education of the jury in the subject matter, such as epidemiology;
3. emphasis on the expert's experience with the disease, injury, or product at issue;
4. explanation of the link between the case at bar and the expert's expertise;
5. education of the jury with respect to the appropriate methodology;
6. education of the jury with respect to the peer review process;

7. education of the jury regarding existing scientific literature relevant to the disease, injury, or product at issue;
8. establishment of the opposing party's views as a misinterpretation of the evidence or an interpretation of the evidence that is premised on unreliable or incomplete evidence;
9. refutation of the opposing party's expert's testimony where it has been speculative; and
10. establishment of the opposing party's expert's nonacknowledgement or nonuse of contrary scientific evidence.

The following excerpt is from the direct testimony of an epidemiologist-expert testifying in a lawsuit against the manufacturers of Bendectin. In this lawsuit, plaintiffs were seeking damages for birth defects in their children which were alleged to have resulted from the mothers' ingestion of Bendectin during the course of their pregnancies. This excerpt contains questions that are typically posed to educate the jury about the discipline known as epidemiology and to educate the jury about the current state of the scientific literature.

> Q: What briefly...is an epidemiologist?
> A: An epidemiologist is someone who studies the pattern of diseases within a group and tries to determine what are the various factors that increase the risk of someone contracting a certain disease... .
> Q: Doctor, I believe that there are basically three areas that you discuss in your report...They include the number of malformations and what you refer to as the "window of opportunity" and the number of children with malformations. Would you please describe what you found and summarize what is contained in the report?

Unlike direct examination, which is conducted by the attorney for the party for whom the expert is appearing, cross-examination is conducted by the attorney for the opposing party. While direct testimony seeks to establish the epidemiologist-expert as an expert and to underscore his or her competence and ability to evaluate the case at bar, cross-examination seeks to cast doubt on his or her ability or opinion. Cross-examination frequently focuses on the limitations of the expert's expertise, the methodology utilized by the expert, the nature of the epidemiological evidence, possible alternative causes of the plaintiff's disease or injury, and any differential diagnoses. Cross-examination may attempt to discredit the expert on the basis of poor memory, a conflict of interest, or lack of qualifications as an expert (Voke, 1994).

Assessing the Qualifications of the Epidemiologist as an Expert

Even though an epidemiologist may have used a methodology that the court finds is reliable, the individual may or may not qualify to testify as an expert witness. Various factors are considered in determining whether an individual should be deemed to be an expert in his or her field, including the individual's education in the specific area; whether he or she has performed or supervised any research in the specific area of inquiry; whether he or she has published any articles in the particular field; and the extent of the prospective witness' familiarity with the scientific literature that addresses the issue at hand, such as the relationship between a particular exposure and a specified disease outcome.

Courts have not infrequently been suspicious of expert witnesses and have questioned the motives that underlie their testimony. One trial court characterized the testimony of a proffered expert as that

> of a crank, or, what is more likely, of a man who is making career out of testifying for plaintiffs in automobile accident cases in which a door may have opened; at the time of trial he was involved in 10 such cases. His testimony illustrates the age-old problem of expert witnesses who are often the mere paid advocates or partisans of those who employ or pay them, as much so the attorney who conducts the suit. There is hardly anything, not palpably absurd on its face, that cannot now be proved by some so-called "experts!" (J. Posner, dissenting, *Chaulk v. Volkswagen of America*, 1986: 808 F.2d at 644).

One circuit court cautioned lower courts to

> be wary lest the expert become nothing more than an advocate of policy before the jury. Stated more directly, the trial judge ought to insist that a proffered expert bring to the jury more than the lawyers can offer in an argument... .
>
> Trial judges must be sensitive to the qualifications of persons claiming to be expert... . [T]he signals of competence cannot be catalogued. Nevertheless, there are almost always signs of both competence and of the contribution such experts can make to a clear presentation of the dispute... . [W]e point by way of example to two. First, many experts are members of the academic community ... . We know from our judicial experience that many such able persons present studies and express opinions that they might not be willing to express in an article submitted to a refereed journal of their discipline or in other contexts subject to peer review. We think that this is one important signal, along with many others, that ought to be considered in deciding whether to accept expert testimony. Second, the professional expert is now commonplace... .
>
> Experts whose opinions are available to the highest bidder have no place testifying in a court of law, before a jury, and with the imprimatur of the trial judge's decision that he is an "expert" (*In re Air Crash Disaster at New Orleans, La.*, 1986: 1233–34).

Yet another judge commented:

> [A]n expert can be found to testify to the truth of almost any factual theory, no matter how frivolous, thus validating the case sufficiently to avoid summary judgment and forcing the matter to trial. At the trial itself, an expert's testimony can be used to obfuscate what would otherwise be a simple case. The most tenuous factual bases are sufficient to produce firm opinions to a high degree of "medical (or other expert) probability" or even of "certainty." Juries and judges can be, and sometimes are, misled by the expert-for hire (Weinstein, 1986: 482).

Even a federal task force has criticized courts' reliance on experts:

> It has become all too common for "experts" or studies" on the fringes of or even well beyond the outer parameters of mainstream scientific or medical views to be presented to juries as valid evidence from which conclusions may be drawn. The use of such invalid scientific evidence (commonly referred to as "junk science") has resulted in findings of causation which simply cannot be justified or

understood from the standpoint of the current state of credible scientific and medical knowledge (United States Attorney General, 1986: 35). (See below for a discussion of "junk science.")

Despite these criticisms and doubts, studies have demonstrated that jurors are not unduly swayed by expert testimony. Jurors not infrequently dismiss experts as "hired guns" (Special Committee, 1940) or downplay or disregard expert testimony in instances where it is particularly complex (Diamond Casper, 1992; Goodman, Greene, Loftus, 1985).

EPIDEMIOLOGICAL CAUSATION MEETS LEGAL CAUSATION

Reconciling Law and Epidemiology

In the context of toxic torts, the plaintiff must establish that the alleged injury arose from the conduct of the defendant. The plaintiff is required to demonstrate that but for the defendant's conduct, the injury would not have occurred. Alternatively, some jurisdictions permit the plaintiff to rely on the "substantial factor test," under which the plaintiff must demonstrate that the defendant's conduct was a substantial factor in bringing about the harm that the plaintiff suffered (Keeton et al., 1994). It is important to recognize that if there is no association between the putative exposure and the alleged disease or injury, there cannot be an inquiry into whether or not the injury to the plaintiff arose from the conduct of the defendant. If at an *in limine* hearing it is determined that the proffered evidence relating Substance A to Disease B is unreliable, that evidence is inadmissible at trial. If that was the only evidence available to the plaintiff to demonstrate the relationship between plaintiff's injury and defendant's conduct, then there is no basis left on which to proceed. For instance, suppose that a plaintiff claims that his stomach cancer resulted from extensive and prolonged exposure to aspirin. Unless he can establish that an association exists between aspirin usage and stomach cancer, his case cannot go forward to determine whether the injury arose from the conduct of the defendant aspirin manufacturer, such as the provision of an inadequate warning with respect to the dangers of extensive aspirin exposure.

In attempting to reconcile the operationalization of causation in an epidemiological context with that used in a legal context, some courts have held that evidence will not be admissible unless it is statistically significant at a 95 percent confidence level, and have equated the scientific standard with the legal standard (Poole, 1987). This interpretation ignores the fact that the significance level is chosen arbitrarily. Other courts have erroneously equated the magnitude of the relative risk with statistical significance. For instance, the court in *In re Joint Eastern & Southern Asbestos Litigation* (1993) held that "a [relative risk] of less than 1.50 is statistically insignificant." Several courts, however, have properly recognized that the absence of statistical significance does not foreclose the existence of a causal relationship between the exposure of interest and the alleged resulting injury (*Allen v. United States*, 1984; *In re TMI Litigation Cases Consolidated II*, 1996).

Some writers have argued that because the legal system requires that the probability of causation have exceeded 50 percent, a person's disease "more likely than not" would have been caused by the alleged exposure only if the relative risk is equal to or greater than 2 or the etiologic fraction is greater than 50 percent (Muscat & Huncharek, 1989). This view has been adopted by some courts. For instance, the court in *Marder v. G.D. Searle & Co.* (1986: 1092) found that

> a two-fold increase is an important showing for plaintiffs to make because it is the equivalent of the required legal burden of proof—a showing of causation by the preponderance of the evidence or, in other words, a probability of greater than 50 percent. In epidemiological terms, a figure of 1.0 indicates no change in the risk, and a figure of 2.0 indicates a two-fold risk.

Other writers have taken an even more restrictive approach, arguing that the courts should require a minimum relative risk of 3.0 to establish causation (Black, Jacobson, Madeira, & See, 1997). This restrictive approach is problematic for several reasons.

First, strict adherence to a risk ratio of 2.0 or more ignores many of the issues inherent in any epidemiological study that affect the magnitude of the resulting measure of association, including sample size and bias. Second, as a matter of policy, the adoption of such a threshold requirement is ill-advised because it would *a priori* preclude recovery by individuals actually harmed by a specific exposure, where less than a relative risk of 2.0 is demonstrated.

This discussion of a threshold requirement based upon the magnitude of relative risk or attributable fraction can be analogized to the epidemiological concepts of sensitivity and specificity (Loue, 1999), used in conjunction with screening tests. Sensitivity refers to the proportion of individuals who truly have a specific characteristic that is correctly classified by the screening strategy as having that characteristic. Specificity refers to the proportion of individuals who do not have a specific characteristic, that is correctly classified by the screening strategy as not having that characteristic. Sensitivity and specificity are related in that attempts to reduce the proportion of individuals incorrectly classified as having a characteristic (false positives) may result in an increase in the proportion of individuals misclassified as not having the characteristic (false negatives) (Roht, Selwyn, Holguin, & Christensen, 1982).

Accordingly, in the context of litigation, a high threshold requirement, such as a relative risk of 3.0, would result in a lesser proportion of unharmed individuals being erroneously classified as harmed, but would also result in a larger proportion of truly harmed individuals being misclassified as unharmed, thereby precluding recovery for those injuries. Conversely, formal adoption of a lower relative risk as a threshold requirement would result in misclassification of a greater proportion of the unharmed as harmed. See Figures 16 and 17, below.

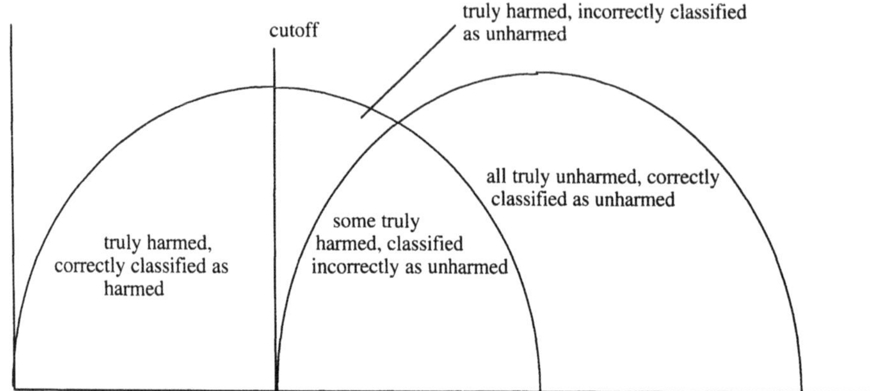

Figure 1-16 Standards for Determining Causation: High Standard

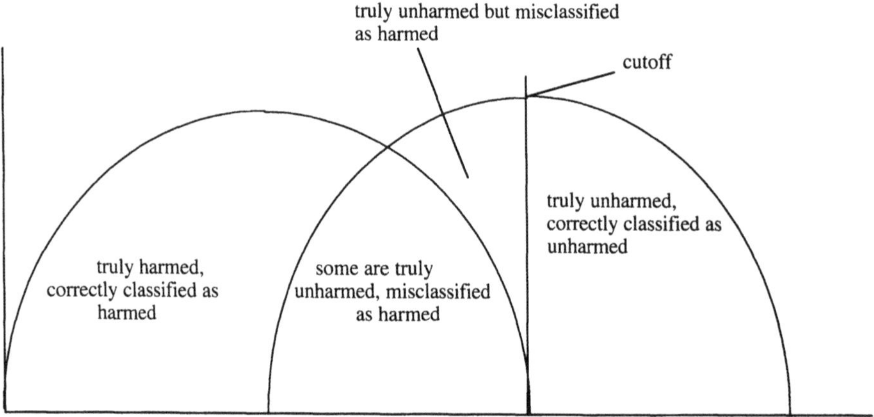

Figure 1-17 Standards for Determining Causation: Low Standard

The establishment of a threshold requirement can also be likened to the issue of Type I and Type II errors. Assume, for instance, that the null hypothesis (H_0) states that a risk ratio of less than or equal to 2.0 indicates no causal association in the legal and epidemiological contexts, while the alternative hypothesis (H_A) states that a risk ratio of greater than 2.0 indicates a causal association exists. In any specific situation, one of the following must be true:

1. The null hypothesis is rejected when it is true and, as a result, a Type I error is committed. This means, for instance, that a jury may conclude that a causal association exists and damages should be awarded when, in fact, there is no such causal relationship.
2. The null hypothesis is rejected when the alternative hypothesis is true, i.e. a correct decision has been made.
3. The null hypothesis is not rejected when the alternative hypothesis is true. This constitutes a Type II error. Practically, this means that the

jury may conclude that there is no causal association and no damages should be awarded when, in fact, such a relationship does exist.

4. The null hypothesis is not rejected when the null hypothesis is true.

Both Type I and Type II errors are serious. A Type I error would result in an award to an allegedly injured party when in actuality the putative exposure is not causally associated with the disease or injury. A Type II error would result in the denial of an award to an individual whose injury or disease may be associated with the exposure under examination.

Whether a Type I or Type II error can be considered more serious in this context rests on a policy determination. It is critical, in this regard, to note that as a society we have recognized and accepted numerous other situations in which compensation schemes fail to reflect a perfect synchronicity between injury and compensation. Such is the case, for example, in the contexts of both workman's compensation for employment-related injuries and no-fault automobile insurance (Macklin, 1999). And, the resolution of this policy determination necessarily rests, at least in part, on how we resolve the underlying ethical dilemma: Is it more unjust to fail to compensate individuals whose exposures did, in fact, result in injury, or more unjust to collect damages from a company for unwarranted liability? (see Macklin, 1999).

Third, it has been argued that the requirement of a relative risk of 2.0 to establish that an individual's injury was "more likely than not" brought about by the alleged exposure collapses the burden of proof with the burden of persuasion. As previously indicated, the burden of proof refers to the facts that the plaintiff must prove in order to establish a *prima facie* case, such as duty, breach of duty, and harm, among others. The burden of persuasion refers to the level of confidence that the jury must have in order to find a fact true for the party that has the burden of proof to prove that fact (Gold, 1986). There are three different burdens of persuasion. "By a preponderance of the evidence" is the standard that is applied in most civil cases, although the standard of "clear and convincing" is also applicable to civil cases. In criminal cases, the burden of persuasion is that of "beyond a reasonable doubt." By collapsing the burden of proof and the burden of persuasion, the standard of proof may be lowered from that of true versus false to that of a probability greater than 50 percent. The burden of persuasion, though, may be heightened (Gold, 1986).

Gold (1986: 382–383) uses the following example to illustrate this difference:

> A traffic light fails to turn red and a crash ensues. The city, sued for its defective light, argues that the car could not have stopped in time even had the light worked, so the light did not cause the accident.

Using a traditional analysis, under which the plaintiff bears both the burden of proof and the burden of persuasion, the plaintiff must prove first that the car could have stopped in time (burden of proof). After finding that the light truly caused the accident, the jury must then evaluate whether the plaintiff has met his burden of persuasion, *i.e.* that causation was more likely than not, meaning that the jury is more than 50% confident in this fact.

Gold's (1986: 382–383) alternative interpretation of the car accident illustrates the collapsing of the burden of proof and the burden of persuasion. In this scenario, we have no information available with respect to the individual car that was involved in the accident. The jury finds that

> We accept the undisputed fact that 53% of cars could have stopped. With no reason to find your car atypical, we infer that your car more likely than not would have stopped, since most cars would have.

Gold (1986) has recommended that the "substantial factor" test be utilized as the burden of proof. This standard recognizes that there may be multiple factors involved in causation and meshes well with Rothman's conceptualization of multiple-factor disease causation. Exposures resulting in a substantially increased risk of disease would consequently result in liability, even if other factors were also implicated in producing the outcome. And, rather than relying solely on a specific test statistic, such as a risk ratio of 2 or an attributable fraction of 50 percent, it is suggested that a determination of causality consider the totality of the circumstances.

The court in *Landrigan v. Celotex Corp.* (1992: 1087) arrived at a similar conclusion, finding that

> a relative risk of 2.0 is not so much a password to a finding of causation as one piece of evidence, among others, for the court to consider in determining whether the expert has employed a sound methodology in reaching his or her conclusion.

Accordingly, the court granted recovery for injuries alleged to have arisen as the result of exposure to asbestos, although the demonstrated relative risk was 1.5.

Epidemiology, Ethics, and Expert Witnesses

Questions have frequently been raised regarding the ethics of an epidemiologist appearing as an expert witness in the context of litigation. A perception exists that, to the extent that participation in a trial as an expert witness constitutes advocacy, such a role "is incompatible with scientific objectivity and impartiality (Last, 1996: 60).

This view, however, is not universal among either epidemiologists or ethicists. One noted ethicist has argued that an inappropriate advocacy role

> may lead an investigator to report results in a fragmentary or unbalanced fashion that promotes the desired values and conclusions ... causing a loss of scientific objectivity. Value-directed roles do not themselves produce impartiality, however. *A partisan or institutionally loyal epidemiologist may restrain or even eliminate his or her beliefs when conducting a study or reporting its findings* (emphasis added) (Beauchamp, 1996: 35).

Others may feel that payment for an epidemiologist's services may result in biased testimony. One scholar has argued that the procurement of experts is best left to the opposing attorneys and that "it is the fact that a scientist holds to a position, and can articulate that position authoritatively, clearly and convincingly that causes him to be sought after as an expert witness and to be well paid (Cole, 1991: 37S). Rather, it is only by looking at the expert's underlying motives and their import on the witness' testimony that one can determine whether a bias actually exists. Those motives may include prestige, power, remuneration, advocacy of truth, professional satisfaction, fear of illness or death, fear of financial loss, a desire for financial retribution, or social, political or economic advocacy (Cole, 1991). Indeed, "to imply, without evidence, that a witness is unethical merely because he harbors one or more of these motives, including the seeking of remuneration, is itself unethical" (Rothman, quoted in Cole, 1991: 38S).

DISCUSSION QUESTIONS

1. Rule 706 of the Federal Rules of Evidence permits a trial judge to appoint an independent scientific expert to help him or her assess the quality of the expert scientific evidence that has been proffered by the parties to the litigation. However, the majority of judges do not utilize this mechanism. What are the advantages and disadvantages to a judge of relying on an independent scientific expert? What are the advantages and disadvantages to the litigants if a judge exercises this option?
2. In some instances, claims that injuries have resulted from specified exposures have been dismissed as having no scientific basis, and then we have found, later, that a causal association actually did exist between the actual exposure and the alleged harm. This occurred, for instance, with exposures claimed to be responsible for the Gulf War syndrome. Is our current system of litigation adequate to address such situations? Why or why not? What modifications, if any, should be made to the system and why?
3. Assume that a man and a woman recently terminated their romantic-sexual relationship. The woman has since discovered that she is HIV-positive. She has sued her former lover, alleging that he knowingly exposed her to HIV.
   a. What are the epidemiological issues involved in establishing the transmission of HIV between these former partners?
   b. What are the legal issues involved in establishing causation?
   c. What do you believe should be the outcome of this lawsuit and why? If you cannot reach a decision, what additional information would you need, if any, to make a determination?

REFERENCES

Allen v. United States (1984), 588 F. Supp. 247 (D. Utah), *reversed on other grounds*, 816 F.2d 1417 (10th Cir. 1987), *cert. denied*, 484 U.S. 1004 (1988).

Anderson v. Owens-Corning Fiberglass Corp. (1991). 810 P.2d 549.

Beauchamp, T. L. (1996). Moral foundations. In S. S. Coughlin, T. L. Beauchamp (Eds.). *Ethics and Epidemiology* (pp. 24–52). New York: Oxford University Press.

Black, B., Jacobson, J. A., Madeira, E. W. Jr., and See, A. (1997). Guide to epidemiology. In B. Black and P. W. Lee (Eds.), *Expert Evidence: A Practitioner's Guide to Law, Science, and the FJC Manual* (pp. 73–115). St. Paul, Minnesota: West Group.

Carrigan, J. R. (1995). Junk science and junk research. *Trial Lawyers Guide, 39*, 230–254.

Carter, K. C. (1985). Koch's postulates in relation to the work of Jacob Henle and Edwin Klebs. *Medical History, 29*, 353–374, 356–357.

Chaulk v. Volkswagen of America, Inc. (1986). 808 F.2d 639 (7th Cir.).

Checkoway, H., & Demers, D. A. (1994). Occupational case-control studies. *Epidemiologic Reviews, 16*, 151–162.

Cole, P. (1991). The epidemiologist as an expert witness. *Journal of Clinical Epidemiology, 44*,, supp. 1, 35S–39S.

Comstock, G. W. (1994). Evaluating vaccination effectiveness and vaccine efficacy by means of case control studies. *Epidemiologic Reviews, 16*, 77–89.

Cook, R. R. (1994). Transcript of videotaped deposition, June 14, in *Ruth Johnson et al. v. Dow Corning et al.*, No. 254–93, District Court, Johnson County, Texas.

Copeland, K. T., Checkoway, H., McMichael, A. J., & Holbrook, R. H. (1977). Bias due to misclassification in the estimation of relative risk. *American Journal of Epidemiology, 105*, 488–495.

Coughlin, S. S. (1990). Recall bias in epidemiologic studies. *Journal of Clinical Epidemiology, 43*, 87–91.

Daubert v. Merrell-Dow Pharmaceuticals, Inc. (1993). 509 U.S. 579, 113 S.Ct. 2786, 125 L.Ed.2d 469.

Daubert v. Merrell-Dow Pharmaceuticals, Inc. (1995). 43 F.23 1311 (9th Cir.).

Diamond, S., & Casper, J. (1992). Blindfolding the jury to verdict consequences: Damages, experts, and the civil jury. *Law and Society Review, 26*, 513–563.

Dwyer, D. M., Strickler, H., Goodman, R. A., & Armenian, H. K. (1994). Use of case-control studies in outbreak investigations, *Epidemiologic Reviews, 16*, 109–123.

Farrar, W. B. (1991). Clinical trials: Access and reimbursement. *Cancer, 67*, 1779–1782.

Federal Rules of Civil Procedure 26, 56.

Federal Rules of Evidence 401, 403, 702.

Felson, D. T. (1992). Bias in meta-analytic research. *Journal of Clinical Epidemiology, 45*, 885–892.

Fitzpatrick, J. M. & Shainwald, S. (1996). *Breast Implant Litigation*. New York: Law Journal Seminars-Press.

Gold, S. (1986), Causation in toxic torts: The burdens of proof, standards of persuasion, and statistical evidence. *Yale Law Journal, 96*, 376–402.

Gore, S. M. (1981). Assessing clinical trials—why randomise? *British Medical Journal, 282*, 1958–1960.

Gray-Donald, K., & Kramer, M. S. (1988). Causality inference in observational vs. experimental studies: An empirical comparison, *American Journal of Epidemiology, 127*, 885–892.

Greenland, S. (1977). Response and follow-up bias in cohort studies. *American Journal of Epidemiology, 106*, 184–187.

Greenland, S., & Robins, J. M. (1985). Confounding and misclassification. *American Journal of Epidemiology, 122*, 495–506.

Hammond, E. C., Selikoff, I. J., & Seidman, H. (1979). Asbestos exposure, cigarette smoking, and death rates. *Annals of the New York Academy of Science, 330*, 473–490.

Hill, A. B. (1965). The environment and disease: Association or causation? *Proceedings of the Royal Society of Medicine, 58*, 295–300.

Hills, M., & Armitage, P. (1979). The two-period cross-over clinical trial. *British Journal of Clinical Pharmacology, 8*, 7–20.

Hymowitz v. Eli Lilly & Co., 73 N.Y.2d 487, *cert. denied*, 493 U.S. 944 (1989).

Glantz, S. A., Slade, J., Bero, L. A., Hanauer, P., and Barnes, D. E. (1996). *The Cigarette Papers*. Berkeley, California: University of California Press.

In re Agent Orange Products Liability Litigation (1984), 597 F. Supp. 740 (E.D.N.Y.), 603 F. Supp. 239 (E.D.N.Y. 1985); 611 F. Supp. 1223 (E.D.N.Y.), *affirmed* 818 F.2d 187 (2d Cir. 1987).

In re Air Crash Disater at New Orleans, La (1986), 795 F.2d 1230 (5th Cir.).

In re Bendectin Litigation (1988), 857 F.2d 290 (6th Cir.), *cert. denied*, 1989 US LEXIS 168.

In re Joint Eastern and Southern Asbestos Litigation (1993), 827 F. Supp. 1014 (S.D.N.Y.), *reversed*, 52 F.3d 1124 (2d Cir. 1995).

In re Silicone Gel Breast Implant Products Liability Litigation (1994). Master File No. CV 92-P-10000-S, interrogatories reprinted in S.M. Mackauf. (1994). *Breast Implant Litigation: Current Medical and Legal Theories.* New York: New York Law Publishing Company.

In re TMI Litigation Cases Consolidated II (1996). 922 F. Supp. 997 (M.D. Pa.).

Jooste, P. L., Yach, D., Steenkamp, H. J., Botha, J. L., & Rossouw, J. E. (1990). Drop-out and newcomer bias in a community cardiovascular follow-up study. *International Journal of Epidemiology, 19*, 284–289.

Karns, E. (1994). Understanding epidemiological evidence: What to ask the expert witness. *Trial*, 30, 48.

Keeton, P. et al. (1984). *Prosser and Keeton on the Law of Torts*, 5th ed. St. Paul, Minnesota: West Publishing.

Kehm v. Procter and Gamble Manufacturing Company (1983), 724 F.2d 613 (8th Cir.).

Kelsey, J. L., Thompson, W. D., & Evans, A. S. (1986). *Methods in Observational Epidemiology.* New York: Oxford University Press.

Khlat, M. (1994). Use of case-control methods for indirect estimation in demography. *Epidemiologic Reviews, 16*, 124–133.

Khoury, M. J., & Terri H., Beaty, T. H. (1994). Applications of the case-control method in genetic epidemiology. *Epidemiologic Reviews, 16*, 134–150.

Kleinbaum, D. G., Kupper, L. L., & Morgenstern, M. (1982). *Epidemiologic Research: Principles and Quantitative Methods.* New York: Van Nostrand Reinhold.

Landrigan v. Celotex Corp. (1992), 605 A.2d 1079 (N.J.).

Lasky, T., & Stolley, P. D. (1994). "Selection of cases and controls," *Epidemiologic Reviews, 16*, 6–17.

Last, J. M., ed. (1988). *A Dictionary of Epidemiology*, 2nd ed. New York: Oxford University Press.

Liberati, A. (1995). "Meta-analysis: statistical alchemy for the 21st century": Discussion. A plea for a more balanced view of meta-analysis and systematic overviews of the effect of health care interventions. *Journal of Clinical Epidemiology, 48*, 81–86.

Loue, S. (1999). *Forensic Epidemiology: A Comprehensive Guide for Legal and Epidemiology Professionals.* Carbondale, Illinois: Southern Illinois University Press.

Lubin, J. H., & Gail, M. H. (1984). Biased selection of controls for case-control analyses of cohort studies, *Biometrics, 40*, 63–75.

Macklin, R. (1999). Ethics, epidemiology, and law: The case of silicone breast implants. *American Journal of Public Health, 89*, 487–489.

Marcus, R. L., & Rowe, T. D., Jr. (1994). *Gilbert Law Summaries: Civil Procedure.* Chicago: Harcourt Brace Legal and Professional Publications, Inc.

Marder v. G. D. Searle and Co. (1986), 630 F. Supp. 1087 (D. Md.), *affirmed* 814 F.2d 655 (4th Cir. 1987).

Matson, J. V. (1994). *Effective Expert Witnessing*, 2d ed. Boca Raton, Florida: Lewis Publishers.

Mauro v. Raymark Industries, Inc., 561 A.2d 257 (N.J. 1989).

Miettinen, O. S. (1985). The case-control study: Valid selection of subjects. *Journal of Chronic Disease, 38*, 543–548.

Miles, & J. Evans (Eds.), *Demystifying Social Statistics* (pp. 87–109). London: Pluto Press.

Morabia, A. (1991). On the origin of Hill's causal criteria. *Epidemiology, 2*, 367–369.

Morgenstern, H. (1996, Spring). *Course Materials, Part II: Class Notes for Epidemiologic Methods II, Epidemiology 201B*, 54–82.

Morgenstern, H. (1982). Uses of ecologic analysis in epidemiologic research. *American Journal of Public Health, 72*, 1336–1344.

Muscat, J. E., Huncharek, M. S. (1989). Causation and disease: Biomedical science in toxic tort litigation. *Journal of Occupational Medicine*, 31, 997–1002.
Novick, J. (1987). Use of epidemiological studies to prove legal causation: Aspirin and Reye's syndrome, a case in point. *Tort and Insurance Law Journal*, 23, 536–557.
O'Brien, P.C., & Shampo, M. A. (1988). Statistical considerations for performing multiple tests in a single experiment. 5. Comparing two therapies with respect to several endpoints. *Mayo Clinic Proceedings, 63*, 1140–1143.
Olkin, I. (1995). Statistical and theoretical considerations in meta-analysis, quoting the National Library of Medicine, *Journal of Clinical Epidemiology, 48*, 133–146.
Poole, C. (1987). Beyond the confidence interval. *American Journal of Public Health, 77*, 195–199.
Popper, K. R. (1968). *The Logic of Scientific Discovery*, 3rd ed. rev. New York: Harper and Row.
Roht, L. H., Selwyn, B. J., Holguin, A. H., & Christensen, B. L. (1982), *Principles of Epidemiology: A Self-Teaching Guide*. New York: Academic Press, Inc.
Rosner, F. (1987). The ethics of randomized clinical trials. *American Journal of Medicine, 82*, 283–290.
Rothman, K. J. (1976). Causes, *American Journal of Epidemiology, 104*, 587–592.
Rothman, K. J. (1986). *Modern Epidemiology*. Boston: Little, Brown and Company.
Sackett, D. L. (1979). Bias in analytic research. *Journal of Chronic Diseases, 32*, 51–63.
Sanders, J. (1992), The Bendectin litigation: A case study in the life cycle of mass torts. *Hastings Law Journal*, 43, 301–418.
Sartwell, P. E. (1960). On the methodology of investigations of etiologic factors in chronic disease further comments. *Journal of Chronic Disease, 11*, 61–63.
Schlesselman, J. J. (1982). *Case Control Studies*. New York: Oxford University Press.
Schuck, P. H. (1987). *Agent Orange on Trial: Mass Toxic Disasters in the Courts*. Cambridge, Massachusetts: Belknap Press.
Schwartz, T. M. (1988). The role of federal safety regulations in products liability actions. *Vanderbilt Law Review*, 41, 1121–1169.
Selby, J. V. (1994). Case-control evaluations of treatment and program efficacy. *Epidemiologic Reviews, 16*, 90–101.
Senn, S. J. (1991). Falsification and clinical trials. *Statistics in Medicine, 10*, 1679–1692.
Shapiro, A. K., & Shapiro, E. (1997). *The Powerful Placebo: From Ancient Priest to Modern Physician*. Baltimore, Maryland: Johns Hopkins University.
Sindell v. Abbott Laboratories, 26 Cal. 3d 588, *cert. denied*, 449 U.S. 912, 1980.
Smith, P.G. (1983). Issues in the design of case-control studies: Matching and interaction effects. *Tijdschrift voor Sociale Gezondheidszorg, 61*, 755–760.
Special Committee on Jury Comprehension of the American Bar Association Section of Litigation. (1989). *Jury Comprehension in Complex Cases*. Chicago: American Bar Association.
Sterling, T. D., Weinkam, J. J., & Weinkam, J. L. (1990). The sick person effect. *Journal of Clinical Epidemiology, 43*, 141–151.
Susser, M. (1986). The logic of Sir Karl Popper and the practice of epidemiology. *American Journal of Epidemiology, 124*, 711–718.
Susser, M. (1991). What is a cause and how do we know one? A grammar for pragmatic epidemiology. *American Journal of Epidemiology, 133*, 635–648.
Swan, S.S. (1994). Epidemiology of silicone-related disease. *Seminars in Arthritis and Rheumatism*, 24, 38–43.
United States Attorney General. (1986). *Report of the Tort Policy Working Group on the Causes, Extent, and Policy Implications of the Current Crisis in Insurance Availability and Affordability*.
Voke, B.P. (1994). Sources of proof of causation in toxic tort cases. *Defense Counsel Journal, 61*, 45–50.
Wacholder, S., Silverman, D. T., McLaughlin, J. K., & Mandel, J. S. (1992). Selection of controls in case control studies. III. Design options. *American Journal of Epidemiology, 135*, 1042–1050.

Weed, D. L. (1986). On the logic of causal inference. *American Journal of Epidemiology, 123*, 965–979.
Weinstein, J. B. (1986). Improving expert testimony. *University of Richmond Law Review, 20*, 473–497.
Weiss, N. S. (1994). Application of the case-control method in the evaluation of screening, *Epidemiologic Reviews, 16*, 102–108.
Windsor Shirt Company v. New Jersey National Bank (1992), 793 F. Supp. 589 (E.D. Pa.).
21 Code of Federal Regulations § 312.21(a)(1998).
21 Code of Federal Regulations § 312.21(b)(1998).
21 Code of Federal Regulations § 312.21(c)(1998).

CHAPTER 2

Case Study One

The Silicone Breast Implant Litigation

> "Here" cried Alice, quite forgetting in the flurry of the moment how large she had grown in the last few minutes, and she jumped up in such a hurry that she tipped over the jury box with the edge of her skirt, upsetting all the jurymen onto the heads of the crowd below, and there they lay spreading about, reminding her very much of a globe of goldfish she had accidentally upset the week before.
>
> "Oh, I *beg* your pardon! She exclaimed in a tone of great dismay, and began picking them up as quickly as she could, for the accident of the goldfish kept running in her head and she had a vague sort of idea that they must be collected at once and put back onto the jury box, or they would die.
>
> "The trial cannot proceed," said the King in a very grave voice, "until all the jurymen are back in their proper places-all," he repeated with great emphasis, looking hard at Alice as he said so.
>
> Alice looked at the jury box and saw that in her haste she had put the Lizard in head downward, and the poor little thing was waving its tail about in a melancholy way, being quite unable to move. She soon got it out again, and put it right. "Not that it signifies much," she said to herself; "I should think it would be *quite* as much use in the trial one way up as the other" (Carroll, quoted in Robbins, 1990: 194).

The dissonance in purpose and the divergence in methodology is well-illustrated by reference to the litigation surrounding injuries alleged to have arisen from the use of silicone breast implants.

THE HISTORY OF SILICONE BREAST IMPLANTS

Silicone, often modified with cottonseed oil or other types of oil, was first used to enlarge breasts among Japanese women in the 1940s and among Las Vegas showgirls in the 1950s (Anderson, 1990). The oil was used in an effort to induce scarring, which would prevent the silicone from migrating to other parts of the body. Approximately 50,000 women in the United States received silicone breast injections, which sometimes resulted in serious injury, including death. Silicone injections were first classified as a drug by the Food and Drug Administration (FDA) in 1965. The FDA has never approved silicone injections for sale for human use.

Silicone gel breast prostheses first became available in the 1960s. It was thought that the use of a silicone gel in lieu of a silicone liquid would prevent the silicone from migrating to other parts of the body. These implants became available prior to the promulgation of FDA regulations requiring safety and efficacy data for medical devices.

The 1976 Medical Device Amendments to the Food, Drug and Cosmetic Act provided the FDA with the authority to regulate breast implants (Public Law 94-295; 21 United States Code §360(c)). The FDA was required, pursuant to this law, to classify medical devices into one of three categories based on the risk associated with the use of the device. Those devices representing the highest risk were classifiable as Class III devices and would require evidence as to their safety and effectiveness. Silicone breast implants were not regulated prior to 1976.

The FDA did not require manufacturers to provide evidence of the safety and effectiveness of the silicone breast implants, although silicone injections were classified as a Class III medical device. A 1978 FDA advisory panel suggested that silicone breast implants be classified as a Class II device, which would not require proof of safety and effectiveness (Federal Register, 1982: 2820). Ultimately, however, the FDA classified both silicone and saline breast implants as Class III medical devices (Federal Register, 1988: 23, 856–23, 877). The classification of the breast implants as a Class III device gave the FDA authority to require that manufacturers submit premarket approval applications to support claims of safety and efficacy, but only after the FDA had issued the final regulation regarding the classification of the implants. (See chapter 4 for a discussion of the use of epidemiology in formulating regulations.) The issuance of the regulation did not occur until April 1991.

Congressional investigators, however, had been receiving reports of problems resulting from the use of silicone breast implants since the 1970s. Dr. Pierre Blais (1990:41), a researcher with the Canadian Department of National Health and Welfare had concluded based on his studies that

> [T]he constituent materials are ill-chosen. Physiologically, and in terms of engineering, they do not reflect the knowledge of our times. The testing that is done on them over the last three decades is trivial, if not totally irrelevant. The performance is far below that of products used in other medical areas ... Laboratory work on collected prostheses indicates a safe lifetime of less than 4 years for many types of prostheses. We are recovering [explanted] prostheses or fragments thereof where the shell and gel are chemically changed. Shells are weak like wet paper. You can tear them easily. Even if they are superficially intact at the moment of explantation, they cannot sustain the capsulotomy, or any type of medical procedure to reduce contracture or to obtain biopsies. The device is finished. To top it off, we have found something else. The tissue around it ... forms an abrasive substance, a material like sandpaper which will ensure the demise of the prosthesis well within the 5-year limit.

In November, 1991, the General and Plastic Surgery Devices Panel of the FDA concluded that the data that had been submitted for its review did not establish the safety of the silicone breast implants (Nightingale, 1992). The FDA announced on April 16, 1992 that silicone gel implants were to be available only through controlled clinical studies (Kessler, 1992). This announcement followed a brief period during which time the FDA had urged the observance of a voluntary moratorium on the use of the silicone implants. The FDA estimated that, during the interval from 1962 until the discontinuation of the device's marketing in 1992, millions of women in the United States had received the implants. A mail survey of approximately 40,000 households found that 8.08 per 1,000 women in the United States reported ever having had some type of breast implant, resulting in a significantly lower estimate of the number of women exposed to implants (Cook & Perkins, 1996; Cook, Delongchamp, Woodbury, Perkins, & Harrison, & 1995). Approximately 20 percent of the devices had been used for reconstruction following mastectomy, with the majority of the devices implanted for breast augmentation (Department of Health and Human Services, 1991).

The General and Plastic Surgery Devices Panel of the FDA had expressed concern regarding a number of safety issues, including the rate of rupture and leakage, the life expectancy of the implants, complications from scar tissue surrounding the implant (capsular contracture), possible interference of the device with mammography, a possible increased risk of malignancy, and a possible association with various immune disorders (*FDA Medical Bulletin*, 1992). Several additional concerns were reflected in internal documents released in 1992 by Dow Corning, one of the primary manufacturers of the silicone gel implant. These documents revealed that women had received breast implants before lifetime tests had been conducted in animals. Preliminary studies had indicated that silicone could migrate or cause other problems. A 1978 internal memorandum of the Medical Engineering Corporation, the breast implant company that was later sold to Bristol-Myers Squibb, had described studies of beagles in which adverse reactions to breast implants included hemorrhage, possible pneumonia, and hyperplasia of the lymphoid tissue in the large intestine. The president of the company had subsequently ordered that the dogs be sacrificed immediately and that their organs be disposed of; the research findings were not to be disclosed publicly (Human Resources and Intergovernmental Relations Subcommittee of the Committee on Government Operations, 1993).

Capsular contracture has been one of the most widely acknowledged problems resulting from the use of silicone breast implants. This occurs when the implant becomes surrounded by a protective layer of scar tissue inside the body. The exact cause of this phenomenon is unknown, but some researchers believe that it represents the body's normal response to a foreign body. Others, however, attribute its occurrence to bleeding, infection, and/or silicone leakage. If the scar tissue shrinks around the implant, it may cause the breast to become harder, more painful, and sometimes misshapened. Contracture can occur anywhere from weeks to years following implantation of the device. Once contracture occurs, the rate of recurrence is high. Contracture can be corrected through either open or closed capsulotomy.

Open capsulotomy is performed as a surgical procedure and, as such, requires the removal of the tissue capsule or the replacement of the entire implant and capsule. Closed capsulotomy requires that the surgeon squeeze the hardened breast by hand; this procedure may result in the rupture of the implant with consequent leakage (Human Resources and Intergovernmental Relations Subcommittee of the Committee on Government Operations, 1993).

Dow Corning had been notified as early as 1982 by University of Chicago researchers that the body reacted to silicone by creating macrophages that eroded the silicone envelope and migrated to the lymph nodes. One researcher indicated that capsular contracture could be the result of the body's immune reaction (Parsons, 1982).

Other problems had surfaced. By 1988, it had been reported that from 22 to 83 percent of glandular tissue was obscured by breast implants, so that cancer screening by mammography became much more difficult (Hayes, Jr., Vandergrist, and Diner, 1988). The compression of the breast that was required for mammography was difficult due to the breast's hardness resulting from contracture. The implant could hide a tumor or make it more difficult to detect changes indicative of early carcinoma (Reynolds, 1995).

EPIDEMIOLOGICAL FINDINGS AND LEGAL VERDICTS

The Events

Numerous epidemiologic studies were conducted in order to investigate the relationship between silicone gel implants and the risk of cancer. One study reported an increased risk of lung cancer among the women who had received implants (Deapen and Brody, 1992). However, that study utilized a relatively small sample and the investigators failed to control for a number of factors, including cigarette smoking and infection with human papilloma virus (Silverman, Brown, Bright, Kaczmarek, Arrowsmith-Lowe, and Kessler, 1996). A study conducted in Alberta, Canada concluded that breast implants neither increased nor decreased the risk of breast cancer, regardless of whether an induction period of zero, five, or ten years was used in the analysis (Bryant and Brasher, 1995).

Many investigations examined the postulated association between silicone breast implants and autoimmune disease. The term "human adjuvant disease" was once used to refer to the constitutional symptoms that were reported by patients who had received implants. However, the term was discarded due to its imprecision. Many of the earlier investigations consisted of case reports or case series, which were characterized by numerous methodological weaknesses: the absence of a clearly defined case definition, the failure to apply conventional diagnostic criteria, the failure to identify the type of implant involved, lack of information relating to the patient's status prior to implant, and/or a failure to utilize a consistent latency period (Kurland & Homburger, 1996). Later investigations utilized case-control and cohort study designs. The following excerpts from a number of these studies are representative of the findings of the research relating to silicone breast implants and connective tissue disease:

> The low incidence of immune diseases in women with breast implants so far reported in the literature and the other cases presented up to now have not provided sufficient data to establish a direct cause-effect relationship between autoimmune disease and silicone implants (Hirmand, Latenta, & Hoffman, 1993: 17).
>
> The causal relationship between augmentation mammoplasty and connective tissue disease remain to be established (Chang, 1993: 469).
>
> Although case reports abound, no convincing evidence exists to indicate that the presence of these two events, autoimmune disease and silicone gel implants, is anything but coincidental. This study did not demonstrate any cause-effect relationship between autoimmune disease and silicone gel breast implants (Shusterman, Kroll, Reece, Miller, Ainslie, Halabi et al., 1993: 1).
>
> The association between the use of silicone breast implants and the later development of connective tissue disease was reviewed. Data from case reports ..., case series, case-control studies, surveys of plastic surgeons, and cohort studies provided no evidence of an association (Edelman, Grant, & van Os, 1994: 183).
>
> Given the present data, there is inadequate evidence to support a causal association between augmentation mammoplasty with silicone gel-filled breast prostheses and the development of autoimmune connective tissue disease or anti-nuclear antibodies (Peters, Keystone, Snow, Rubin, & Smith, 1994: 1).
>
> We found no evidence of an association between silicone breast implants and either connective tissue diseases defined according to a variety of standardized criteria or signs and symptoms of connective tissue disease (Sanchez-Guerrero Colditz, Karlson, Hunter, Speizer, & Liang, 1995: 1666).
>
> The data from this large retrospective cohort study are compatible with previous recent reports from two other cohort studies that provide reassuring evidence against a large hazard of breast implants on connective tissue diseases (Hennekens., Lee, Cook, Hebert, Karlson, & LaMotte, 1996: 616).
>
> No epidemiologic study has indicated that the rate of well-defined connective tissue disease or breast cancer has greatly increased in women with silicone breast implants.... (Silverman, Brown, Bright, Kaczmarek, Arrowsmith-Lowe, & Kessler, 1996: 1).
>
> The results of the study do not support the hypothesis that silicone gel-filled implants induce or promote CTD [connective tissue disease] (Edworthy, Martin, Barr, Birdsell, Brant, & Fritzler, 1998: 254).

These conclusions regarding the association between silicone breast implants and cancer and silicone breast implants and connective tissue disease can be contrasted with the many verdicts for plaintiffs and the sizable sums awarded to plaintiffs for damages for injuries alleged to have arisen from the use of silicone breast implants. For instance, compensatory damages awarded to plaintiffs have ranged from $1.7 million in the 1984 California case of *Stern v. Dow Corning*, to $25 million in *Johnson v. MEC* (Texas, 1992). Although punitive damages have sometimes been denied, as in *Mahlum v. Dow Chemical* (Nevada, 1995), punitive awards have also been quite high, as in *Johnson v. MEC*, in which $20 million was awarded. A number of lawsuits have also resulted in verdicts for the defendant manufacturers, such as in *Craft v. McGhan* (Hawaii, 1992), *Mohney v. Baxter/Heyer-Schulte* (Colorado, 1993), *Turner McCarthy v. Dow Corning* (Colorado, 1993), *Berry* et al. *v. Baxter* (Texas, 1995), *Habel* et al. *v. Baxter Healthcare* (Texas, 1995), *Hall v. Baxter Healthcare* (California, 1995), *Luevano v. Baxter* (Texas, 1995), *Scott v. Dow Corning* (Texas, 1995), and *Jennings v. Baxter* (Oregon, 1996) (Fitzpatrick, 1996).

Against this backdrop, Dow Corning filed for bankruptcy, proposing in 1998 a $3.2 billion settlement of the claims of approximately 170,000 plaintiffs (Stolberg, 1998). Later that same month, the European Committee on Quality Assurance and Medical Devices in Plastic Surgery concluded that, based on all of the data presented, implants did not cause autoimmune or connective tissue disease. The Independent Review Group created by Great Britain's Minister of Health later reached a similar conclusion, finding that there existed "no histo-pathological or conclusive immunological evidence and no epidemiological evidence" that silicone breast implants caused disease (Bandow, 1998: A23). Later that same year, the expert scientific panel that had been appointed by Judge Samuel Pointer, who oversaw thousands of silicone breast implant cases in federal district court, declared: "The main conclusion that can be drawn from existing studies is that women with silicone breast implants do not display a silicone-induced systemic abnormality in the types or functions of cells in the immune system." It was less clear, the panel noted, as to whether the implants caused "sufficient local inflammation to account for the symptoms women report" (Burton, 1998: B1).

How, one must ask, with such mounting epidemiological evidence against a causal relationship between silicone breast implants and either connective tissue disease or cancer, could such a state of affairs come to pass?

The Explanations: A Post-Mortem

The Cynical View: Junk Science at Its Worst

The more cynical view of how plaintiffs were able to prevail in many of the lawsuits would hold that the courts erroneously permitted reliance on "junk science." This supposedly fatal flaw is deemed to have resulted, at best, in a misunderstanding of the real issues and the data by the jurors and/or judge, or, at worst, the distortion of science and the methods of scientific inquiry to satisfy the rapacious demands of personal injury attorneys with no commitment to Truth or Justice (see Angell, 1996).

The term "junk science" has been used to refer to the introduction of scientific evidence that is not considered to be consistent with accepted scientific views. Huber (1991: 2) has explained that junk science represents

> the mirror image of real science, with much of the same form but none of the same substance. There is the astronomer on one hand and the astrologer on the other. The chemist is compared with the alchemist, the pharmacologist with the homeopathist. Take away the serious sciences of allergy and immunology, brush away the detail and rigor, and you have the junk science of clinical ecology. The

orthopedic surgeon is shadowed by the osteopath, the physical therapist by the chiropractor, the mathematician by the numerologist and cabalist.... Junk science cuts across chemistry and pharmacology, medicine and engineering. It is a hodge-podge of biased data, spurious inference, and logical legerdemain, patched together by researchers whose enthusiasm for discovery far outstrips their skill. It is a catalog of every conceivable kind of error: data dredging, wishful thinking, truculent dogmatism, and now and again, outright fraud.

What was once known as the United States Department of Health Education, and Welfare (1978: xii, quoted in Huber, 1990: 277) had earlier set forth characteristics of individuals believed to engage in diagnostic and therapeutic movements not generally accepted by medical professions—what Huber would refer to as junk science:

> The proponents don the mantle of science while at the same time degrading the reputable scientists of their day;
>
> They claim that the prejudice of medicine hinders their efforts;
>
> They cite examples of physicians and scientists of the past who were forced to fight the rigid dogma of their day;
>
> They rely heavily on testimonials and anecdotes as evidence that their remedy is a safe and effective cancer drug;
>
> They do not use regular channels of communication, such as journals, for reporting scientific information, but rely instead on the mass media and word of mouth;
>
> Their chief supporters are not people trained or experienced with treating cancer or in scientific methodology;
>
> They offer a simplistic scientific theory for causation of the disease;
>
> Their remedy is easy and pleasant, compared with the frightening therapies wielded by orthodox physicians;
>
> And they claim the mode of administration of a drug and method of treatment can be learned only from them.

"Junk science" can be contrasted with what has been called "good science":

> "Good science" is a commonly accepted term used to describe the scientific community's system of quality control which protects the community and those who rely upon it from unsubstantiated scientific analysis. It mandates that each proposition undergo a rigorous trilogy of publication, replication, and verification before it is relied upon (Brief of *New England Journal of Medicine, Journal of the American Medical Association, and Annals of Internals Medicine*, 1993).

Using such definitions, it is clear that numerous scientific theories or tenets might have once been dismissed as junk science because of a lack of consensus or general acceptance within the relevant scientific communities. These include the theory that the earth is round, Darwin's theory of evolution, and the theory that veterans of the Vietnam war might have suffered from posttraumatic stress disorder (Carrigan, 1995). These examples challenge the view of scientific knowledge as developing through a gradual "process of accumulation and... incorporation of knowledge within existing theories" (Trial Lawyers Public Justice Foundation, 1993, cited in Carrigan, 1995: 237) and posit, in contrast, a "reconstruction of prior theory and the re-evaluation of prior fact, an intrinsically revolutionary process...." (Kuhn, 1970: 7).

Kesan (1997) has proposed a temporal model to explain the process inherent in the development of scientific understanding with respect to a specific issue. He posits that scientific methodology follows an evolutionary course that commences with (1) an embryonic phase, followed by a (2) rapidly evolving phase, that culminates in (3) a mature phase. The embryonic phase involves the testing of relationships between two or three variables, based on previous research. The evolving phase is characterized by attempts to disprove the findings of the earlier embryonic phase. The final phase is characterized by the general acceptance of the methodology. The speed with which a methodology progresses from the second, evolving phase, to the third, mature phase, is a function of the perceived significance of the preliminary findings, the specific area of science, and the number of researchers engaged in investigations on the specific topic (Kesan, 1997). The phase of evolution during which an underlying methodology is subject to legal examination may well be related to both a determination regarding the legal reliability and admissibility of the proffered evidence and a determination regarding its sufficiency in establishing a link between the exposure or action of the defendant and the harm that is alleged to have occurred as a result of that exposure or action. It is important to note, too, that Kesan's model is premised on various assumptions: that scientific work progresses openly rather than in secret and that a particular theory or methodology is of interest to numerous researchers. Both of these assumptions are themselves questionable (Bourdieu, 1997).

Juries Achieving Justice

Several legal scholars have posited that,

> [r]ather than acting incompetently by ignoring or even shunning science in entering plaintiff verdicts, jurors may be "commingling" or nullifying the causation rule to produce a legal outcome that compensates for the lack of legal incentives to test products earlier in the development process. According to this theory, jurors combine the weak evidence on causation with the compelling evidence on negligence to support a net finding of liability. This possibility is supported by firsthand accounts of the jury trials and jury deliberations in breast implant cases; the fact that plaintiff verdicts are characterized by a high median and frequency of punitive damage awards; and parallel reports of jury behavior in similar toxic tort cases.... Indeed, juries in the breast implant cases may simply be doing *sub rosa* what the legal system should do formally: shift the burden of proof to manufacturers to disprove causation when the absence of safety research is due to the manufacturer's own neglect (Dresser, Wagner, & Giannelli, 1997: 741–742).

This perspective rests on significant evidence of the implant manufacturers' failure to conduct the requisite research and on their failure, and the failure of many physicians performing breast implant surgery to make prospective patients aware of the potential risks inherent in such procedures. A Congressional Subcommittee Staff report observed that

> in the case of breast implants, there is a 30-year history involving approximately 1 million American women. Although the companies knew since at least 1982 that they would probably be required to provide safety data, and although they were

> warned in 1988 that data would be required in approximately 30 months, many of the studies were not started until 1990 or 1991. Whereas prospective studies that followed women for many years would have been considered ideal, a reasonable alternative would be to start a study in 1990 that asked patients from the 1970's or early 1980's about any medical problems they have had since their implant surgery. That kind of thorough retrospective study was not conducted by any of the manufacturers (Human Resources and Intergovernmental Relations Subcommittee on Government Operations, 1993: 27).

Like the more cynical view, this perspective is critical of our courts. However, rather than holding the courts culpable for their failure to monitor the admission of evidence more adequately, this view implicitly recognizes the failure of our legal system to allocate more fairly the burdens of litigation amongst those who are and who are not responsible for creating the context in which the litigation arises.

An alternative, but similar, view would hold that the plaintiffs' verdicts for the plaintiffs effectuate three of the goals that have been delineated for our tort system: the allocation of resources to those who have been injured by unduly risky conduct or products, the deterrence of excessively risky conduct and the manufacture of excessively risky products, and the compensation by injurers to the victims of their overly risky activities (Feldman, 1995). The enumeration of goals in this manner mirrors the concept of strict products liability, one of the legal theories that plaintiffs argued provided a basis for recovery. The underlying rationale for this theory was first enunciated in 1944:

> Those who suffer injury from defective products are unprepared to meet its consequences. The cost of an injury and the loss of time or health may be an overwhelming misfortune to the person injured, and a needless one, for the risk of injury can be insured by the manufacturer and distributed among the public as a cost of doing business. It is to the public interest to discourage the marketing of products having defects that are a menace to the public. If such products nevertheless find their way into the market it is to the public interest to place the responsibility for whatever injury they may cause upon the manufacturer, who, even if he is not negligent in the manufacture of the product, is responsible for its reaching the market. However intermittently such injuries may occur and however haphazardly they may strike, the risk of their occurrence is a constant risk and a general one. Against such risk there should be general and constant protection and the manufacturer is best situated to afford such protection (*Escola v. Coca Cola Bottling Co.*, Traynor, J., concurring, 1944: 441).

This theory was more explicitly delineated by the California Supreme Court in 1963 as a basis for finding a defendant liable:

> A manufacturer is strictly liable in tort when an article he places on the market, knowing that it is to be used without inspection for defects, proves to have a defect that causes injury to a human being.... The purpose of such liability is to insure that the costs of injury resulting from defective products are borne by the manufacturers that put such products on the market rather than by the injured persons who are powerless to protect themselves (*Greenman v. Yuba Power Products, Inc.*, 1963: 900–901).

The defect may take one of three forms: a defect in the manufacture of the item, a defect in design, and/or a defect in the warning provided by the manufacturer in conjunction with the marketing or use of the product. Manufacturing defects are unintended failures of the quality control process. A design defect indicates that the product should have been formulated in a safer manner, rather than the manner used. A defect in the warning reflects inadequacies in the information provided by the manufacturer in the information provided regarding the risks that attend the usage of the product. Many of the lawsuits that were filed for injuries alleged to have arisen from the use of silicone breast implants centered on the manufacturers' failure to warn plaintiffs of the risks that attended the use of their product. For example, a complaint alleging failure to warn in the context of strict liability might have stated:

1. Defendants are corporations whose agents, servants, and employees designed, manufactured, sold, and distributed the breast implants that are the subject of this litigation.
2. The breast implants manufactured and supplied by defendants failed to comply with representations of defendants that they were fit for implantation into the human body.
3. The breast implants manufactured and supplied by defendants were defective in design and manufacture in that they failed to function properly.
4. The breast implants manufactured and supplied by defendants were defective in design or manufacture in that when they left the hands of defendants, the foreseeable risks exceeded the benefits associated with their design or manufacture.
5. Alternatively, the breast implants manufactured and supplied by defendants were defective in design or manufacture in that, when they left the hands of the manufacturer and/or suppliers, they were more dangerous than an ordinary consumer would expect.
6. The breast implants manufactured and supplied by defendants were defective due to inadequate warning or instruction because the manufacturer knew or should have known that the product created a risk of harm and defendant failed to warn of the risk.
7. The breast implants manufactured and supplied by defendants were defective due to inadequate postmarketing warning or instruction because, after the defendants knew or should have known of the risk of harm from the defective product, the defendants failed to provide an adequate warning to implant recipients.
8. As a direct and proximate result of the defective breast implants, plaintiffs suffered and will suffer harm and economic loss.

A failure to warn may also be alleged in the context of a claim of negligence. To establish liability for a failure to warn under this theory, the plaintiffs must demonstrate that "a manufacturer or distributor did not warn of a particular risk for reasons which fell below the acceptable standard of care, i.e., what a reasonably prudent manufacturer would have known or warned about" (*Anderson v. Owens-Corning Fiberglass Corp.*, 1991: 558). Neither the standard of care nor the manufacturer's conduct are relevant in the context of a claim for failure to warn using a theory of strict liability. A failure of the manufacturer to warn of the risks, in essence, vitiates informed consent to the implantation of the product because an individual cannot make an informed decision regarding whether or not to undergo the implantation if the individual is not provided with the information that should form the basis of that decision.

The Cyclical Theory

The silicone breast implant cases can be thought of as, rather than as individual cases, a collection of cases that share certain common characteristics. This "congregation" of cases arises as a result of the mobilization efforts of potential claimants and the attempts of defendants to modify the number and outcome of the lawsuits (Galanter, 1990). Lawyers, courts, and legislatures respond to this phenomenon in a variety of ways. Lawyers may coordinate strategy and share information. Courts and legislatures may attempt to ration recoveries. "Depletion" may occur, as the easiest of the cases are resolved quickly, leaving a smaller and smaller number of cases in which it becomes increasingly difficult for plaintiffs to prevail. Outcome-stabilization may result, in which determinations of liability and the award of damages become more predictable (Galanter, 1990). This "cyclical theory of mass torts" has been explained as follows:

> In the early stages of the cycle, defendants tend to win more cases than plaintiffs because of strategic and informational superiority. If the litigation has any merit, however, plaintiffs will eventually develop successful information and strategies and win an extremely high percentage of the cases tried. Next, the plaintiffs will bring cases for trial that stretch the envelope of viable plaintiffs too far, and defendants will create more effective counter strategies, resulting in a reduced percentage of plaintiff victories. Eventually, after full aggregation and dissemination of information, crystallization of the law, and thorough development of strategies, there will be a rough equilibrium of trial results. Remaining variations will then be due to jury demographics, attorney caliber, and random events during trials. Although perhaps it is counter-intuitive, settlements will also reflect this equilibrium: the average settlement amount will be virtually identical to the average jury verdict. The variance, however, will be substantially different. Settlements for similarly situated plaintiffs will be extremely similar; verdicts will vary in accordance with idiosyncrasies of the trial process (McGovern, 1986: 482).

The existence of such a cycle is clearly evident in the context of the silicone breast implant litigation. The earliest lawsuits were more likely be resolved in defendants' favor. However, as plaintiffs secured increasingly more damaging information related to defendants' conduct, such as the failure to conduct the necessary research and the failure to inform women of the potential risks of implantation, the numbers of plaintiff awards appeared to increase.

The Misguided Jury

Yet another theory would hold that the liability determinations in favor of the plaintiffs, with their attendant awards of damages, reflect the difficulties that jurors experience in their attempts to make sense of the evidence presented and the manner in which it is presented. First, jurors may attempt to distinguish which expert testimony is correct based upon the experts' credentials and testimony (see Shuman, Whitaker, & Champagne, 1994). If jurors are unable to determine which testimony is correct using such criteria, they will attempt to make an evaluation based on personal judgments about the experts, rather than about the information that they present (Goodman, Greene, & Loftus, 1985). One study of experts found that

> communication skills, the ability to convey technical information in a non-technical fashion, and the willingness to draw firm conclusions are what makes them effective in testifying. Physical attractiveness and personality were considered more important for juries than for judges. Educational credentials and being a leading expert in the field were considered less important for juries than judges. Experts thought that a sense of certainty and an ability to communicate not only results in their retention, but also makes them an effective witness (Shuman, Whitaker, & Champagne, 1994: 199–200).

Jurors may erroneously conclude that the "correct" testimony is that which is represented by the most expert witnesses, so that whichever side presents the most experts must be right (Pennington & Hastie, 1992). Jurors may also assume that the view of the expert who testifies in court accurately reflects the consensus of opinion within the relevant professional field:

> In civil litigation ... all manner of experts are found to testify opposite their colleagues.
>
> Whether such "balancing" of expert witnesses helps the fact finder evaluate their testimony is another matter. The search for witnesses that is driven by the adversary process may result in a distortion of knowledge when applied to expert witnesses. For example, if 999 of every 1000 experts in a given field hold one view of a question and one holds an alternate view, the two experts who appear in court will have been detached from the extremely skewed distribution of opinion from which they were drawn. The fact finder has no way of knowing this (Saks & Wissler, 1984: 439–440).

Plaintiffs' success may also be attributable to their ability to construct a believable story. The verdict that results will reflect the consistency between the story that the jurors have constructed in their attempt to understand and to integrate the evidence that has been presented and the jury instructions provided that offer a range of possible legal outcomes (see Pennington & Hastie, 1992). The jury will be influenced, in particular, by the organization of the evidence as a narrative, rather than a compendium of legal issues, and by the ability of the proffered story to address and explain all of the evidence that is presented during the trial (Pennington & Hastie, 1988, 1992). The importance of storytelling has been noted in the context of criminal trials:

> Any competent presentation by the prosecution would form a story. Those jurors favoring the prosecution would have something to relate in a coherent way during deliberations. The strongest counter, therefore, was always another narration, or another way to order the information into coherency, but this time in a way that pointed towards acquittal. Those jurors favoring the defense needed a story if they were going to be successful in persuading other jurors. Thus, at least in my

> experience, the goal of the criminal defense attorney was to process the information into a story indicating the client's innocence, or at least into a story that raised a reasonable doubt about his guilt. Although we could not prove it, we all felt that stories mattered to juries and affected how information was processed (Jonakait, 1991: 347).

In the context of the silicone breast implant litigation, plaintiffs who prevailed were successful in the construction of a story that presented the defendants as callous, greedy corporations, uninterested in and unconcerned with the actual effects of their product, who coldheartedly dismissed the sufferings of consumers of their product in the interest of increased profits. Conversely, losing plaintiffs were at times portrayed by the defense as superficial women with meritless claims, represented by attorneys out to make a fast buck.

DISCUSSION QUESTIONS

1. Significant controversy exists regarding the admissibility of scientific evidence in litigation, with some professionals arguing for the exclusion of "junk science" and others arguing that all expert opinions should be admitted, and the final judgment left to the jury. What are the benefits and risks of each approach?
2. Pharmaceutical companies have maintained that they spend significant sums on research and the development of new and critical products, such as new antibiotics and vaccines, with little insulation from liability for even unforeseeable injuries. One proposal has been the indemnification by the federal government of companies for their development and production of new pharmaceutical products.
   a. What are the potential short- and long-term scientific, economic, and ethical implications of such a plan?
   b. If an indemnification plan were to be adopted, how should guidelines for the compensation of harmed consumers be developed and structured?
   c. What are the advantages and disadvantages of this approach compared to the current system of compensation through litigation, as exemplified by the case of silicone breast implant litigation?

REFERENCES

Anderson, N. (1990). Testimony Before the Subcommittee of the Committee on Government Operations, House of Representatives, *Is the FDA Protecting Patients From the Dangers of Silicone Breast Implants*, Dec. 18, cited in Human Resources and Intergovernmental Relations Subcommittee of the Committee on Government Operations of the House of Representatives. (1993). *The FDA's Regulations of Silicone Breast Implants*, 102nd Congress, 2nd Session, December 1992. Washington, D.C.: Government Printing Office.

Angell, M. (1996). *Science on Trial: The Clash of Medical Evidence and the Law in the Breast Implant Case.* New York: W.W. Norton and Company, Inc.
Bandow, D. (1998). Many torts later, the case against implants collapses. *Wall Street Journal*, November 30, A23.
Blais, P. (1990, Dec. 18). Testimony Before the House Committee on Government Operations, *Is the FDA Protecting Patients from the Dangers of Silicone Breast Implants*, 101st Cong., 2d Sess.
Bourdieu, P. (1997). *Sociaux de la science. Pour une sociologie clinique du champ scientifique.* [Social uses of science. Towards a clinical sociology of the scientific field]. Paris: INRA.
Bryant, H., & Brasher, P. (1995). Breast implants and breast cancer—Reanalysis of a linkage study. *New England Journal of Medicine, 332*, 1535–1539.
Burton, T. M. (1998). Implant makers get a boost from a report. *Wall Street Journal*, December 2, B1.
Carroll, L. *Alice's Adventures in Wonderland.* Quoted in S. Robbins. (1990). *Law: A Treasury of Art and Literature.* New York: Hugh Lauter Levin Associates, Inc.
Chang, Y. H. (1993), Adjuvanticity and arthrogenicity of silicone. *Plastic and Reconstructive Surgery, 92*, 469–473.
Cook, R. R., Delongchamp, R. R., Woodbury, M., Perkins, L. L., & Harrison, M. C. (1995). The prevalence of women with breast implants in the United States—1989. *Journal of Clinical Epidemiology, 48*, 519–525.
Cook, R. R., & Perkins, L. L. (1996). The prevalence of breast implants among women in the United States. *Current Topics in Microbiology and Immunology, 210*, 419–425.
Deapen, D. M., & Brody, G. S. (1992). Augmentation mammoplasty and breast cancer: A five year update of the Los Angeles study. *Plastic and Reconstructive Surgery, 89*, 660–665.
Department of Health and Human Services. (1991). Silicone gel-filled breast prostheses; silicone inflatable breast prostheses: Patient risk information. Federal Register, 56, 49,098–49,101.
Dresser, R. S., Wagner, W. E., & Gaiannelli, P. C. (1997). Breast implants revisited: Beyond Science on Trial. *Wisconsin Law Review, 1997*, 705–776.
Edelman, P., Grant, S., van Os, W. A. (1994). Autoimmune disease following the use of silicone gel-filled breast implants: A review of the clinical literature. *Seminars in Arthritis and Rheumatism, 24*, 183–189.
Edworthy, S. M., Martin, L., Barr, S. G., Birdsell, D. C., Brant, R. F., & Fritzler, M. J. (1998). A clinical study of the relationship between silicone breast implants and connective tissue disease. *Journal of Rheumatology, 25*, 254–260.
Escola v. Coca Cola Bottling Co. (1944). 150 P.2d 436 (Cal.).
Federal Register (1982, Jan. 19). 47, 2810–2853.
Federal Register (1988, June 24). 53: 23,856–23,87.
Feldman, F., Finch, M., & Dowd, B. (1989). The role of health practices in HMO selection bias: A confirmatory study. *Inquiry, 26*, 381–387.
Fitzpatrick, J. M. (1996). MDL update: Breast implant cases. In J. M. Fitzpatrick and S.Shainwald (Eds.)., *Breast Implant Litigation* (pp. 2–6). New York: Law Journal Seminars-Press.
Galanter, M. (1990). Case congregations and their careers. *Law and Society Review, 24*, 371–395.
Goodman, J., Greene, E., & Loftus, E. F. (1985). What confuses jurors in complex cases: Judges and jurors outline problems. *Trial, November*, 65–77.
Greenman v. Yuba Power Products, Inc. (1963). 377 P.2d 897 (Cal.).
Hayes, H., Jr., Vandergrist, J., & Diner, W. C. (1988). Mammography and breast implants. *Plastic and Reconstructive Surgery, 82*, 1–6.
Hennekens, C. H., Lee, I. M., Cook, N. R., Hebert, P. R., Karlson, E. W., & LaMotte, F., et al., (1996). Self-reported breast implants and connective tissue diseases in female health professionals. *Journal of the American Medical Association, 275*, 616–621.
Hirmand, H., Latenta, G. S., & Hoffman, L. A. (1993), Autoimmune disease and silicone breast implants. *Oncology, 7*, 17–24.
Huber, P. (1991). *Galileo's Revenge: Junk Science in the Courtroom.* New York: Basic Books.
Huber, P. (1990). Junk science and the jury. *University of Chicago Legal Forum, 1990*, 273–302.

Human Resources and Intergovernmental Relations Subcommittee of the Committee on Government Operations of the House of Representatives. (1993). *The FDA's Regulations of Silicone Breast Implants*, 102nd Congress, 2nd Session, December 1992. Washington, D.C.: Government Printing Office.

Jonakait, R. N. (1991). Stories, forensic science, and improved verdicts. *Cardozo Law Review, 13*, 343–352.

Kesan A. (1997). A critical examination of the post-*Daubert* scientific evidence landscape. *Food and Drug Law Journal, 52*, 225–251.

Kessler, D. A. (1992). The basis of the FDA's decision on breast implants. *New England Journal of Medicine, 326*, 1713–1715.

Kuhn, T. (1970). *The Structure of Scientific Revolutions* (2nd ed.). Chicago: University of Chicago Press.

Kurland, L. T., & Homburger, H. A. (1996). Epidemiology of autoimmune and immunological diseases in association with silicone implants: Is there an excess of clinical disease or antibody response in population-based or other "controlled" studies? *Current Topics in Microbiology and Immunology, 210*, 427–430.

McGovern, F. (1986). Toward a functional approach for managing complex litigation. *University of Chicago Law Review, 53*, 440–492.

Nightingale, S. L. (1992). Moratorium on silicone gel implants. *Journal of the American Medical Association, 267*, 787.

Parsons, R. (1982, May 14). To G. Jukubczak, Dow Corning, released by Dow Corning February 10, 1992.

Pennington, N., & Hastie, R. (192). Explaining the evidence: Tests of the story model for juror decision making. *Journal of Personality and Social Psychology, 62*, 189–206.

Peters, W., Keystone, E., Snow, K., Rubin, L., & Smith, D. (1994). Is there a relationship between autoantibodies and silicone-gel implants? *Annals of Plastic Surgery, 32*, 1–5.

Public Law 94–295, 90 Stat. 539 (1976), codified at 21 USC § 360.

Reynolds, H. E. (1995). Evaluation of the augmented breast. *Radiologic Clinics of North America, 33*, 2231–1145.

Saks, M. J., & Wissler, R. L. (1984). Legal and psychological bases of expert testimony: Surveys of the law and of jurors. *Behavioral Science and Law, 2*, 435–449.

Sanchez-Guerrero, J., Colditz, G. A., Karlson, E. W., Hunter, D. J., Speizer, F. E., & Liang, M. H. (1995). Silicone breast implants and the risk of connective tissue diseases and symptoms. *New England Journal of Medicine, 332*, 1666–1670.

Shuman, D. W., Whitaker, E., & Champagne, A. (1994) An empirical examination of the use of expert witnesses in the courts—Part II: A three city study. *Jurimetrics Journal, 34*, 193–208.

Shusterman, M. A., Kroll, S. S., Reece, G. P., Miller, M. J., Ainslie, N., & Halabi, S. et al. (1993). Incidence of autoimmune disease in patients after breast reconstruction with silicone gel implants versus autogenous tissue: A preliminary report. *Annals of Plastic Surgery, 31*, 1–6.

Silverman, B. G., Brown, S. L., Bright, R. A., Kaczmarek, R. G., Arrowsmith-Lowe, J. B., & Kessler, D. A. (1996). Reported complications of silicone gel breast implants: An epidemiologic review. *Annals of Internal Medicine, 124*, 744–756.

Stolberg, S. G. (1998). Neutral experts begin studying dispute over breast implants. *New York Times*, July 23, A21.

Trial Lawyers Public Justice Foundation. (1993). Supreme Court rejects "general acceptance" requirement for scientific testimony. Public Justice. Cited in Carrigan, J. R. (1995). Junk science and junk research. *Trial Lawyers Guide, 39*, 230–254.

United States Department of Health, Education, and Welfare. (1978). FDA, Laetrile—The Commissioner's Decision [HEW Pub. No. 77-3056], *quoted in* Huber, H. W. (1990). Junk science and the jury. University of Chicago Legal Forum, 273–302.

CHAPTER 3

Case Study Two

The E. Coli Investigation

> Article 4. We forbid all emptying or tossing out into the streets and squares of the aforementioned city and its surroundings of refuse, offals, or putrefactions, as well as all waters whatever their nature, and we command you to delay and retain any and all stagnant and sullied waters and urines inside the confines of your homes. We enjoin you to carry these and promptly empty them into the stream and give them chase with a bucketful of clean water to hasten their course.
>
> Article 15. We forbid all and any persons to leave or dispose of any manner of fodder, animal wastes, soot and ashes, mud or any other kind of unspeakable waste on the streets. Nor may these streets be used for conflagrations or the slaughter of pigs or beasts of any kind. Indeed, we enjoin such persons to collect droppings and wastes and to gather them inside their homes, where they shall pack them into receptacles and wicker baskets to be then carried outside the aforementioned city and its surrounding areas (Paris edict, November 1539, quoted in Laporte, 2000: 4–5).

Consider the following scenario. The county health department in a rural area of a state is notified that two children have been hospitalized at the local hospital due to bloody diarrhea. One of the children has a confirmed culture of *E. coli* O157:H7. On the same day, a physician in the same community notified the health department of two additional children suffering from bloody diarrhea. Ultimately, the parents of these children, and others, sued the school district for damages arising from what they alleged was the school's mishandling of lunch meat served to children at the elementary school. Was there, in fact, a causal association between the meat and the bloody diarrhea as an epidemiologist would define it? Was legal causation established? How would you have handled the investigation of the putative link?

THE EPIDEMIOLOGY OF *E. COLI*

E. coli is a common inhabitant of the intestinal tract of both humans and warm-blooded animals. It was, for a long time, regarded as essentially harmless. However, outbreaks of foodborne occurring in November 1971 were found to have been caused by *E. coli* contained in Camembert and Brie cheeses imported from France (United States Department of Health, Education, and Welfare, 1971).

Serotypes of *E.coli* that can cause foodborne gastroenteritis are known as enteropathogenic. There are four major serogroups of enteropathogenic *E. coli*: (1) classical sergroups associated with diarrhea in young children and infants; (2) serogroups associated with sporadic diarrhea and normal flora in the intestinal tract; (3) enterotoxigenic E. coli associated with traveler's diarrhea; and (4) strains known as enteroinvasive because they can result in the invasive infection of the gastrointestinal tract (Kornacki and Marth, 1982).

The enterotoxigenic type of E. coli has a mean incubation period of 26 hours, compared to a mean incubation period of the invasive type of just 11 hours. The enterotoxigenic type is characterized by diarrhea, vomiting, dehydration, and shock, all symptoms that are also associated with cholera. The invasive type of E. coli has as its primary symptoms abdominal cramps, watery stools, fever, chills, and headache.

E. coli flourishes best in a temperature of 98.6 degrees Fahrenheit, with a range in temperature from 50 to 104 degrees. The ideal pH is close to neutral, but *E. coli* can survive in a range of pH 4 to 8. Contamination with *E. coli* has been found to originate from many sources due to its ubiquitous nature: water, dust, air, food, kitchen utensils, rodents, flies, and food handlers. Adequate cooking temperatures and pasteurization are the usual means of preventing infection.

THE FACTS

Finley is a small town that is located in a rural area of Benton County in Washington State. Private wells supply water to residences, irrigation systems, and family farms and pastures. The school district includes a high school, consisting of grades 9 through 12, a middle school for grades 6 through 8, and an elementary school for those in pre-school through the fifth grade. The elementary school alone employs 55 people and serves 466 students.

Each of the schools serves breakfast and lunch. Meals are usually prepared in a central kitchen at the high school and are then transported to each of the other schools for additional cooking and preparation prior to serving. Water is provided to the elementary school through a private well system that is inspected on a regular basis by the health district. Sewage disposal is accomplished via a private drain field.

On October 16, 1998, the school district sent a letter to all parents advising them that two of three children confirmed to have become ill as the result of infections with *E. coli* O157:H7 had been attending the elementary school. The letter further advised the parents that the health district had determined that the common element in two cases was the children's ingestion of taco meat at the elementary school on October 6. The third student indicated, however, that she had not eaten the taco meat; the health department had no explanation for the cause of her infection. Parents were advised of the safety precautions that were being taken to prevent additional infections. The letter advised parents to seek medical attention for their children if they were experiencing any of the following symptoms: abdominal pain, cramps, mild fever, painful bloody diarrhea after four days of illness, nausea, watery diarrhea, and vomiting (Van Slyke, 1998a). A subsequent letter was sent on October 20, which again advised parents that the taco meat was suspected of having harbored the bacteria and that all necessary and appropriate measures were being taken to avoid further transmission of the infection.

THE EPIDEMIOLOGICAL INVESTIGATION

The health department conducted a multi-pronged investigation of the outbreak, which it characterized as parent interviews, a case-control study of the students, a cohort study of the students, and a cohort study of the school staff.

The Initial Investigation

The initial investigation consisted of interviews with the parents of patients at the office of a local physician and at the hospital. The health department report stated that all cases had attended a school in the Finley school district. The health department determined that the exposure to the bacteria must have occurred during the week of October 5 through 9 and, because some of the children attended school only on Tuesdays, and Thursdays, it was likely that the exposure had actually occurred on either Tuesday, October 6, when taco meat was served, or on Thursday, October 8, when a ham and cheese sandwich was served.

In attempting to pinpoint the likely time of exposure, the health department constructed four different epidemiologic curves. None of the curves bore a date or time, so that an individual unfamiliar with the progression of the investigation would be unable to ascertain which of these curves represented the actual data and which had been discarded. These curves appear below as curves A through D.

The Case-Control Study

The health department conducted a case-control study to assess the risk of contracting the infection among those who had been exposed to the taco

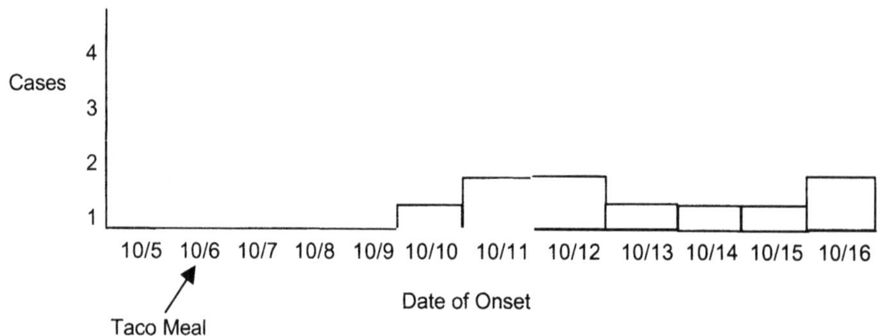

Figure 3-18 Curve A. Epi Curve By Onset of Diarrhea, Confirmed and Probable *E. coli* O157—Kennewick: Finley School

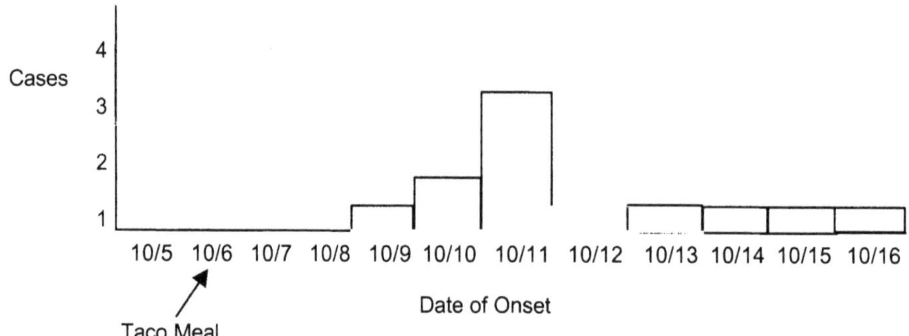

Figure 3-19 Curve B. Epi Curve By Onset of Illness, Confirmed and Probable *E. coli* O157-Kennewick: Finley School

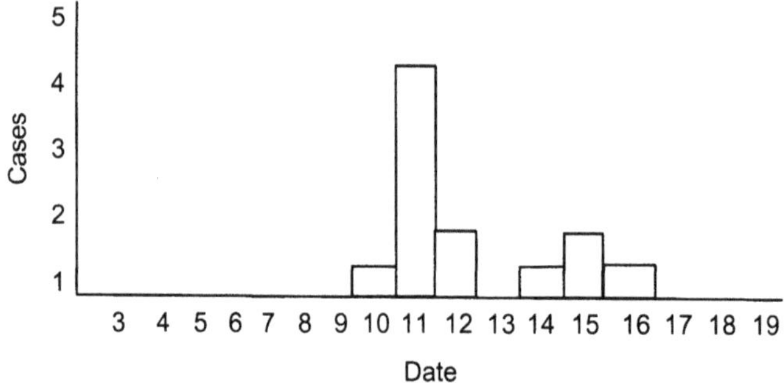

Figure 3-20 Curve C. Number of Confirmed and Probable Cases *E. coli* O157:H7 by Onset of Illness, Finley Elementary School, October 1998

meat and those who had not. A confirmed case of *E. coli* was defined as a resident of or visitor to Finley with culture-confirmed *E. coli* O157:H7 diarrhea that occurred after October 1, 1998 and with epidemiologic linkage to Finley Elementary School. A probable case was defined as a resident of or

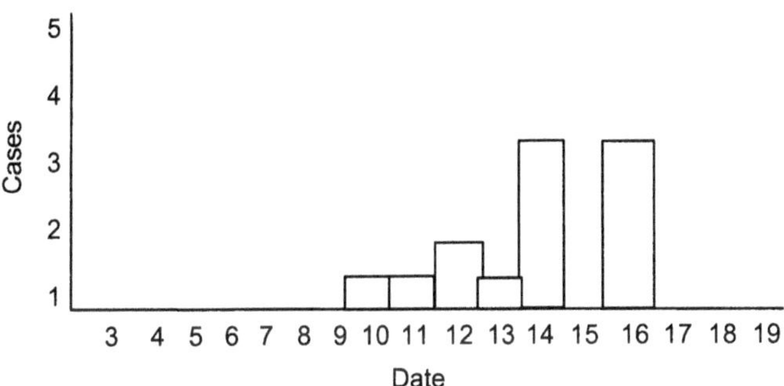

Figure 3-21 Curve D. Number of Confirmed and Probable Cases *E. coli* O157:H7 by Onset of Illness Finley Elementary School, October 1998

visitor to Finley who experienced bloody diarrhea of unknown origin after October 1, 1998, and/or complications of hemolytic uremic syndrome, with epidemiologic linkage to Finley Elementary School. Controls consisted of classmates of the cases, matched by classroom. Data from the cases were collected through in-person interviews, while control were interviewed by telephone. The health department stated that data were collected on each child's age, grade, ingestion of menu items at the school cafeteria, contact with livestock, ingestion of food outside of the school setting, exposure to water, contact with children in diapers, and travel.

The Student Cohort Study

The health department collected data on the order in which different classrooms ate their lunches in the school cafeteria.

The Staff Cohort Study

Staff and adult volunteers at Finley Elementary School were asked to complete a self-administered questionnaire. These forms included items relating to illnesses during the first three weeks of October, the treatment received for the illnesses, and the consumption of meals that were prepared at the school.

Findings

The health department ultimately found 9 confirmed cases of *E. coli* O157:H7 and 2 probable cases. There were cases in all grades of Finley Elementary School with the exception of grade 4. The mean age of the infected children was 8 years and the mode was 10 years. No statistical association was found between ingestion of any food item and illness. The epidemiologist for the health department constructed the tables 3-1 & 3-2 from the data. Fisher's exact test, 2-tailed was used to calculate the odds ratios and confidence intervals for both tables.

The response to the staff survey was moderate, with 34 of 55, or 62%, of the staff and volunteers responding. Five of them reported having had diarrhea, three of whom had diarrhea within the definitional time frame. One of the staff having diarrhea had eaten both the taco meat and the ham sandwich meal from the student lunch line. Two other adult staff reported having eaten the taco meal but they did not report any illness and they did not report having had their food from the student line.

Environmental investigation revealed that the ground beef used for the taco meat was from frozen stock and had been prepared in one large batch in a steam jacketed kettle at the central kitchen. The meat was then portioned out and transported to the various schools. The van that delivered the beef to

Table 3-1 Odds Ratio of Illness Associated with Foods Served at FES, October 5–8, 1998

Food Item	# Cases Eating Item	# Controls Eating Item	Odds Ratio (95% CI)	p-Value
Chicken Nuggets	9	15	4.20 (0.38, 108.22)	0.38
Dipping Sauce	5	9	1.81 (0.29, 11.61)	0.69
Milk (10/5)	8	15	1.87 (0.24, 17.09)	0.68
Taco, Beef	9	15	4.20 (0.38, 108.22)	0.38
Taco, Cheese/ Lettuce	6	10	1.50 (0.25, 9.33)	0.70
Taco, Tomatoes	3	5	1.50 (0.19, 11.50)	0.67
Buttered Corn	7	9	2.59 (0.40, 18.43)	0.43
Milk (10/6)	9	18	1.42 (0.10, 41.64)	1.0
Hot Dog	9	13	2.77 (0.38, 24.80)	0.43
Tomato Soup	5	4	3.54 (0.54, 24.86)	0.21
Veggie Sticks	5	6	1.94 (0.33, 11.96)	0.45
Ham Sandwich	7	11	1.91 (0.29, 13.25)	0.69
Celery Sticks	2	4	1.21 (0.12, 11.30)	0.76
Pears	7	14	1.75 (0.22, 16.48)	0.68

Table 3-2 Relative Risk of Illness Associated with Meals Served at FEs, October 5–8, 1998

Meal	# Ill Eating Meal	# Not Ill Eating Meal	Relative Risk (95% CI)	p-Value
Chicken	9	15	4.20	0.38
Bkfst 10/5	0	75	0.00	0.13
Lnch 10/5	8	262	1.22 (0.33, 4.54)	1.00
Bkfast 10/6	1	89	0.34 (0.04, 2.60)	0.46
Lnch 10/6	9	286	1.51 (0.33, 6.87)	0.73
Bkfst 10/7	0	77	0.00	0.13
Lnch 10/7	8	260	1.25 (0.34, 4.65)	1.00
Bkfst 10/8	1	71	0.45 (0.06, 3.44)	0.69
Lnch 10/8	9	194	1.35 (0.30, 6.14)	1.00

the elementary school did not have hot holding capacity. The beef that was served at the elementary school on the day in question was stored in warming bins prior to being served. At the other schools, the meat was placed in trays on the stove top and burners were lit to keep the food warm. The ham and cheese sandwiches in questions had been prepared by heating them in the warming bins, whereas at the other schools, they were prepared in regular ovens. Water was supplied to the elementary school through a private well system. Sewage disposal was effectuated through a private drain field.

Laboratory investigation of the remaining cooked meat and samples of the remaining frozen meat resulted in negative test findings for *E. coli* O157:H7. Water samples tested negative for fecal coliforms.

Tests were conducted on isolates from confirmed case patients. The subtype of *E. coli* was compared to the subtype from previous outbreaks, but it did not match any subtypes from recent cases in either the Washington State or CDC database. This presumably included an outbreak of *E. coli* that had occurred the previous summer, the source of which was never identified.

Although the children had a common play period, the health department concluded that as

> no other common school activity was identified other than eating at the school cafeteria, it is reasonable to conclude that a meal served at the school was the likely source of illness. Cattle are the known reservoir of *E. coli* O157:H7. Thus, it is likely that consuming the ground beef served in the tacos was the vehicle.
>
> Findings supporting this conclusion are:
>
> a. The confirmed and probable cases were restricted to the elementary school.
> b. Results of the PFGE [pulsed-field gel electrophoresis] analysis of E. coli O157:H7 isolates from confirmed case patients indicate that an identical subtype of E. coli infected the confirmed cases.

c. No confirmed or probable cases were reported from other schools in the district.
d. No teachers or staff reported illness to BFHD [Benton-Franklin Health District] or WDOH [Washington Department of Health]. Most staff did not eat food supplied from the student serving line.
e. No confirmed or probable cases were reported from the community at large which matched the FES [Finley Elementary School] subtype.
f. Results of the environmental investigation suggest that although similar menu items and common food stock were served to all schools in the district, there were differences in food preparation technique, transportation time, and hot holding techniques between the elementary school and the other schools in the district. In particular, deficiencies in the transport method had been noted in previous inspections.
g. Results of the case-control study suggest that the cases did not have common risky exposures outside of the elementary school.
h. Results of the epidemiologic studies did not reveal any special school activities (birthday parties, field trips, etc.) that could account for possible exposures during this time period...

Ground beef is a known vehicle for *E.coli* O157:H7 and this investigation noted differences in the preparation, handling, and transport of meat which may have allowed for uneven cooking, uneven cooling, and uneven reheating at the elementary school. This outbreak and the resulting investigation highlight the importance of regular inspections of institutional kitchens and the need for ongoing training of food service workers.

The report noted several limitations, including the failure to validate self-reports of food consumption or non-consumption in the cafeteria against the school lunch payment records, the possibility of recall bias, and the possibility of having overmatched in the selection of controls for the case-control study.

DISCOVERY

Plaintiffs' Case

Plaintiffs' attorneys wanted to show that *but for* the ingestion of the allegedly contaminated ground beef in the taco meat, the children with *E. coli* would not have otherwise become infected. This strategy is evident in the questions that were posed to the expert witnesses and their various responses. Epidemiologists for the defense (the school system) are designated as Epidemiologist D1 and Epidemiologist D2. The two epidemiologists hired by the attorneys for the plaintiffs (the parents acting on behalf of the children who became sick) are designated as Epidemiologist P1 and P2.

Epidemiologist D1 had a Ph.D. in microbiology and was employed as a clinical epidemiologist for an acute and communicable disease program of a state health department. The second defense expert, Epidemiologist D2, had a Ph.D. in epidemiology and was an epidemiologist on faculty at a well-known medical school. Excerpts from their depositions appear below.

From the deposition of Epidemiologist D1 by plaintiffs' attorneys:

Q: Does it remain your view that the overwhelming majority of O157 infections, if traced back to their ultimate source, will end up at a bovine source?
A: How would you define overwhelming majority?
Q: Well, something over 60 percent.
A: Yes.
Q: How about over 75 percent?
A: Quite possibly.
Q: Okay. And if it's not bovine, then your view would be that there are other mammalian sources for this reservoir?
A: Some people use the word reservoir to mean any animal that has been shown to have it; and if that's what you mean, I certainly think that there are other mammalian-potential mammalian and possibly non-mammalian sources for human infections.
Q: What would those primary sources be?
A: Deer and other cervines. Elk, in particular, in our experience have been sources of infection.

Other animals—a number of other animals have been known to carry the organism and be possible sources for cases around the world—horses, dogs, sheep,—although none of those have been identified as more than the possible or probable source of a sporadic case or two to date.

Defendant's Case

Attorneys for the defense maintained that there exist other mechanisms for the transmission of *E. coli* and that other potential exposures had not been investigated during the course of the investigation. This strategy is reflected in the questions posed of plaintiffs' experts.

From the deposition of Epidemiologist P1 by defendant's attorneys:

Q: You said that you can't identify a particular food source with O157:H7 because you said it crosses different sources. What did you mean by that?
A: It could exist in cows. It can also infect humans. It exists in deer and it can infect humans.
Q: Is there a limitation in the animal world as to where O157:H7 can exist or do we know?
A: We don't know. It hasn't been extensively studied. Right now we have hit or miss. You know, just we find it [sic] in different animals just by accident because we are looking but there hasn't been any large scale studies to look at all these animals and see first if you can find it there. If you find it there, are they just passing through the system, they just picked up the contamination. And you sample them two more weeks—two weeks later they are not going to be there or are they in co-existence with this animals and you can routinely, you know, sample the same cat or the same duck or the same dog and isolate the organism.
Q: How about with cattle? Do you know if there is either a parasitic or a symbiotic relationship between E. coli and cattle or are they just passing through?

A: Well, there have been studies. People at Washington State University have done the most work on this. And some cows you know, they test consistently positive and you can test them after six months and they have it and some other ones they just clear it. It's there for a month and then it's not there. But usually it kind of persists in the herd. You know, once you have two, three hundred cows in one farm, you know, at any given time if you do a lot of samplings likely that one of them or a few of them are infected and then they're just-it just passes around.

Q: Is O157:H7 something that is endemic to the cow or do they pick it up from an external source or do we know?

A: I don't think we really know that.

Q: How is O157:H7 transmitted generally?

A: From?

Q: Well, let me break it down. Let's see. We know that O157:H7 can be contained in ground beef and then ingested by humans. That is a method of transportation of O157. Would you agree?

A: That's one method.

Q: Can it be transmitted in other ways, for example, human contact?

A: Yes, it could.

Q: What's the mechanism of that?

A: It's called fecal oral route. Person who is infected they can get massive numbers of the organism on their hands, they can deposit it. With kids I mean they always have one hand, you know, then another one in their friend's mouth. This is called secondary transmission. Direct contamination of people or contamination of surfaces, contamination of foods that people handle.

Q: How about if one walks through a herd of cows and steps in the fecal matter that's on the ground? Is there a possibility of transmittal of O157:H7 by that method?

A: If you are saying that can they pick it up-if fecal matter is contaminated, can they pick it up on their shoes, for example, and then take it to home and then, you know, your dog comes in and licks their feet and then licks their face later on.

Q: Sure.

A: Yes, it's conceivable.

Q: Well, is it a —when you say conceivable—

A: Yes.

Q: —is it reasonable that if there is O157:H7 in fecal matter and somebody puts their hand in it, for example, that they would have O157:H7 on their hand?

A: Yeah, they would. Exactly, yes.

Q: How about other animals? Can other animals transport, say, an animal walks through a field or through fecal matter that has O157:H7 can it then be transported by that vehicle, by the animal, to a secondary kind of infection?

A: The way you are describing it is conceivable, sure.

Q: In other words, this is not—this is not a bacterium that would when exposed to air would die, it lives in an aerobic environment?

A: Yes. It would die if it desiccates.

Q: Okay.

A: If it dries it will die....

Q: Okay. How about transportation via insects? Do we know anything about whether or not O157:H7 can be transported through flies or other insects?

A: We haven't had cases that described that route, but depending on the environment if you have a place where you know there is fecal material deposited which contains organisms or insects that go feed there and then they come in and feed on your food then there would be some sort of mechanical transport. Again it's conceivable. You cannot rule this out. These are just very basic principles of tropical medicine. You know, these are all routes of transmission that we know (Samadpour, 2000: 46–50).

Although the plaintiffs' asserted that the *E. coli* infection could only have been caused by exposure to contaminated beef at the Finley Elementary School, the answers given by one of plaintiffs' witnesses, Epidemiologist P1, itself cast doubt on this theory. Epidemiologist P1 was trained as a Ph.D. level microbiologist and had had one course in epidemiology. He was employed as a molecular epidemiologist.

From the deposition of Epidemiologist P1 by defense attorneys:

Q: [B]ased on your earlier testimony am I correct that based on your analysis of the isolates from [the four] children that you cannot state what—that you can state that you believe that they were infected from a single source; is that right?
A: In all likelihood they were exposed to the same source.
Q: More likely than not?
A: Yes.
Q: But you cannot tell us from your work what that source was?
A: Exactly.
Q: And you cannot tell us from your work that [child A] was infected if he was with O157:H7? We don't know that for sure, right?
A: From my work I cannot say that he was not infected or he was infected.
Q: And then going a step further, you certainly can't say then that he was infected from the same source as the other four children?
A: I cannot form an opinion on that.
Q: And with regard to the children for whom isolates were not forwarded to you, other children who are claiming E. coli illnesses, you would not be able to express an opinion as to whether or not they were infected from the same source?
A: If I don't have their isolates I cannot make that, no....
Q: Have you formed an opinion regarding with regard to the children for whom you were not forwarded isolates for evaluation?
A: No.
Q: Have you formed an opinion regarding the source of E. coli infection for the children for whom you evaluated isolates?
A: No.
Q: And have you formed an opinion with regard to the source of the alleged E. coli infection for any of the children for whom you were not forwarded isolates for evaluation?
A: Again, no (Samadpour, 2000: 32–34).

Experts for the defense noted numerous weaknesses in the methodology utilized to conduct the epidemiological investigation, including bias, and arithmetic errors in calculating the odds ratios.

From the deposition of Epidemiologist D1 by plaintiffs' attorneys:

Bias

A: [W]hen they describe some of the history of what happened on Friday, October 16, again—and this is I don't quibble with the account of it, but it's unfortunate they the parents and community and the media were notified about what someone thought was a possible source of the outbreak from the get-go. That would have compromised the ability of any investigator to do an unbiased investigation (Keene, 2000: 37).

Mathematical and Methodological Errors

A: I remember that I—On page 3 there are some—the results of a case control study presented, and specifically in table one there are unmatched odds ratios presented for certain of the food items. I remember I calculated some of these and found some what appeared to be calculation errors in them, although they're not ones that had any substantive effect on the conclusion; but I was just surprised that there appeared to be arithmetic errors (Keene, 2000: 40).

A: The most telling point to me is the distribution of onset times for the cases. The taco meat, as I understand it, was provided only in the lunch on Tuesday, October 6. The typical incubation period for O157—and I'm sorry, and there's nothing in the record to suggest that it was served on Wednesday or Thursday or at any other time.

There are a number of different clocks of onset time, and it's not entirely clear to me which one is the official one. But that said, with the typical incubation period being three to four days, sort of on average three to five days, you would expect the onsets to sort of be distributed around a date that is three to four days after the 6th; and, in fact, they're distributed quite a bit later than that. So that to me is by far the most compelling evidence that points away from the hamburger. It just is pointing away from any exposure at all on Tuesday, the 6th....

Alternative Explanations

A: The distribution of onset times for the cases is not sharp—the case onsets are not sharply clustered.

With a disease like O157 that has a quite variable range of incubation periods, you don't necessarily expect a tightly clustered distribution of onset times, as might be expected for something like Staph aureus or Clostridium perfringens, or even Salmonella, salmonellosis.

So with that caveat, and the lack of other evidence in the record to have a pinpointed source, leaves me at least open to the possibility that exposures occurred over a range of time that was wider than, say, a single meal or an hour or two. But how, you know, it's just a possibility I would keep on the table as an investigator (Keene, 2000: 46–47).

Q: Is it significant to your opinion that the testing done on ground beef after the incident was negative?

A: That is a factor, in my opinion.

Q: Okay, Tell me how big a factor that is.

A: Well, as an epidemiologist, we're very fond of saying that nor finding implicated food or a bug in implicated food doesn't mean that it wasn't there. But when you have the available product, the product available for testing, and it's tested, and you don't find it, it decreases the likelihood that that was the source.

The question is does it decrease it in a trivial—you know, going from 99 percent to 98.9 percent, or is it 90 percent to 5 percent? You can't quantify it. And I don't want to imply that I think that not finding it is devastating to the hypothesis that it was in the ground beef, but it's a factor in my decision (Keene, 2000: 57–58).

The second epidemiologist was deposed by plaintiffs' attorneys particularly on issues related to her assessment of the methodology employed to investigate the putative association between the taco meat and the *E. coli* infections. She commented on various methodological difficulties with the investigation as it had been conducted and as the data had been analyzed:

> There are numerous discrepancies in marking the date and time of onset and the date and time of diarrhea onset, which makes it very difficult to draw an epidemic curve which in turn makes it difficult to figure out where the period of incubation may have occurred. There were other common-there were other sources of ground beef that were common to several of these students that were not investigated .... None of this is investigated ... (Loue, 2000: 16). The problem is to what extent the report of parents and school workers and anyone else was biased because the initial material that was issued into the communities said we think it is ground beef. We don't know that (Loue, 2000: 18–19).
>
> I have one graph that says: Epi curve by onset of diarrhea confirmed and probable E. coli O157 Kennewick Finley school. Then I have another graph that says: Epi curve by onset of illness confirmed and probable E. coli O157 Kennewick Finley school. Then I have another graph that said: Number of confirmed and probable cases E. coli O157:H7 by onset of illness, Finley Elementary School, October 1998. Then I have another graph that says Number of confirmed and probable cases E. coli O157:H7 by onset of diarrhea Finley Elementary School, October 198. Then I have statistical output that indicate that has one frequency distribution marked onset and another frequency distribution marked DIA onset, presumably diarrhea onset. Again, those figures are not the same as the figures that are on four different graphs (Loue, 2000: 21–22).

Even plaintiffs' own expert witness, perhaps unwittingly, found fault with the methodology that had been employed in the effort to identify the source of the children's infection.

From the deposition of Epidemiologist P2 by defense' attorneys:

> Q: Would you agree with me that from an epidemiological standpoint, it would not be a good practice to suspect a particular source and then construct a study directed at that particular source?
>
> A: No. But there are some instances where that wouldn't be true .... So you—it's a narrowing process when you're doing that ... (Alexander, 2000: 13–14).
>
> Q: Have you done an O157:H7 investigation where you had concluded as to the probable source of the O157 when you did not have either an epidemiological link or positive test results?
>
> A: No, I don't think so.
>
> Q: And really from an epidemiological standpoint, you need one or the other in order to make a reasonably scientific conclusion as to the source, don't you?
>
> A: Right (Alexander, 2000: 23–24) ....
>
> Q: Is it good epidemiological practice before you have the results of a case study or control study to presume the source and then commence the investigation with that presumption in mind?
>
> A: It isn't anything that you would want to do—let me say it another way. Can I say it again?
>
> Q: Yes.

A: In public health practice sometimes you end up having to suggest a very likely source before it's absolutely proven. As a matter of fact, sometimes that has caused trouble. It's a big risk on the part of the epidemiologist to have that happen. But—it does prejudice some of the results you get from that point on. But if you had to do it for some other reasons—I think in this case—I don't know what the motivation was for that letter going out very early suggesting that it was that meal... I would think more it potentially might prejudice the results you get from your questionnaires then, because people now think this is the source. So it may have had an effect.... (Alexander, 2000: 50–51).

Q: Did you do any investigation or arrive at any conclusions as to the—whether or not there was any epidemiological link to this outbreak at Finley and the ham and cheese sandwich meal that was served on October 8th?

A: I did not investigations other than to read these reports. The reports—the studies that were done were I think inconclusive, the epidemiological studies that were done, case control and analytical studies that were done, those cohort studies that were done—(Alexander, 2000: 52).

THE TRIAL

Plaintiffs' Case

Experts for the plaintiffs attempted to minimize the methodological weaknesses that had been stressed repeatedly during discovery.

Direct Examination of Epidemiologist P1:

Q: I want to talk a little about the—I think everybody is potentially a little perplexed about this one issue is that how can we point to the school lunch, the taco meat school lunch, as the source of this E. coli 0157:H7 infection if in fact one of now 11 children who became sick didn't eat the school lunch? Can you help us understand how something like that can occur in your opinion?

A: Because this isn't physics or chemistry. This is interviewing humans, and particularly interviewing young humans, children, whose memory of things is not always accurate whose—who may have had—may have as has been brought out earlier today, who may have shared food and so on that might not be admitted to or might not have been fully explored. And it just within my experience it's not unusual to have one or two outliers as it were. Very often if you keep on going back to those outliers, and I can think of a number of things in my experience where suddenly they—they do fall into place. Someone forgot that they did such and such at a certain time. And I don't know how many times whether they—this was done in this instance of going back and requestioning, but it's not unusual to have one outlier like that.

Q: Due to the fact you have what you call the outlier, due to the fact you have this outlier, can you therefore draw the conclusion based on that piece of evidence tat it wasn't the school lunch, taco meat school lunch? Can you draw the conclusion that it wasn't that that caused this illness an this outbreak?

A: No, I wouldn't rule it out on the basis of that one (Alexander, direct testimony, 2001: 605–606).

Interestingly, even the plaintiffs conceded in their presentation of evidence that they could not state conclusively that the ground beef was, in fact, the source of the *E. coli* infection. The second epidemiologist for the plaintiffs testified during direct examination that

> In this instance ... we were not able to unequivocally state that the source was the taco meal on October the 6th. We were unable to use our epidemiologic methods to have a statistically significant result. We were unable to find E. coli in the taco meat.
>
> However, our feeling was that there was sufficient evidence to indicate that while we could not say definitively that the source was 015—the taco meat, we felt that that was very likely the case or the most probable cause was that. In other instances in food-borne outbreaks we're not able to identify a source at all. It's not unusual for that to occur. We investigate about 40 or 50 of these outbreaks of other sorts every year, and it's very common not to identify the particular agent. In this instance we had a very great suspicion that the source was the taco meal (Kobayashi, 2001: 173).

Defendant's Case

The themes that arose during discovery were again raised at the trial itself. Both of the experts for the defense testified that the cause of the epidemic was undetermined and emphasized the methodological problems inherent in the studies relied upon by the plaintiffs' to prove their case.

Defense Direct Examination of Epidemiologist D1:

> Q: What is the primary feature or the evidence or investigation that you believe most significantly draws you to the conclusion that the cause is undetermined as opposed to the taco meat served on October 6th?
>
> A: Well, it's undetermined because none of the traditional measures of—bits of epidemiologic evidence that point to the source really exist, other than this historical fact that ground beef has been in the past one of many—it's the most famous and the ones people in the public hear about the most, but there's very little if anything in this record to point the finger at the ground beef.
>
> Q: As far as the aspects of the investigation, was there one particular feature of the investigation that you looked at as being more persuasive to you that it wasn't thr ground beef or at least you couldn't conclude that it was probably the ground beef on October 6th?
>
> A: The thing that's the most compelling thing pointing away from the ground beef as opposed to just not pointing at anything is the distribution and the incubation—excuse me, the onset times of the children who became ill. They're quite long after the taco meal was—the taco meal was on October 6th, and the distribution of incubation times and onset times just doesn't look very god for exposure on that date.... Well, the incubation period is a key concept for most of these communicable diseases. And for 0157:H7 specifically it's relatively long and variable compared to some of the organisms that cause food-borne disease. It's shorter than some others, but it's typically between two and six or seven days and most cases would average—most groups would average around three or four days from exposure to, um, onset of symptoms.
>
> Now, it could be as short as a day perhaps or as long as eight or ten days in some cases, but those would be exceptions rather than the rule. And if you have a group of people, you can usually look at the average of that group and

> get an idea, pretty good idea, of when the exposure would have most likely took place....
>
> Well, the median is one way of expressing the average. It's the midpoint, and half of the people would be more than this, and half of the people would be less than this. And since there are 11 people, you just put them in order, which they've done on the graph. And all you have to do is find number 6, because there are five more than number 6 and five less than number 6. And number 6 on this graph is occurring on the 14th, three, four, five, six, right? So there are five cases. So you'd say the median incubation date was—excuse me, median onset date was on the 14th....
>
> Well, there's another factor that hasn't been mentioned that's also important to consider, and that's assuming that the exposures all took place at a single time. I think that was the working assumption in this investigation that this all stemmed from a single exposure at one point in time as opposed to exposures over several days or several weeks or something like that. And there are a lot of outbreaks that are like that, but if we assume that this is a single-source exposure point source, then again pointing back from the 14th I'd have to say an exposure around the 9th or so.... (Keene, direct examination, 2001: 1637–1640).

Epidemiologist D1 also referred to the existence of bias:

> A: Well, the bias that I'm referring to, um, stemmed largely from the fact that almost immediately the outbreak was recognized, a statement was made to the public and to everybody that read the newspaper, kinds in the school, that the source was probably the taco meat. So that was kind of what everybody had on their mind from the outset even before a formal investigation was begun really.
>
> Q: And is that something that people in your business or your job and health workers in the local level should try and avoid? Or is that good? What's the problem with that?
>
> A: Well, we would—we wouldn't have been very happy if that happened in our state, yes.
>
> Q: How does it affect the investigation?
>
> A: Well, a lot of what we rely on is our ability to get information from people using pretty simple devices like just making up questionnaires and asking people about questions about events that might not have been all that memorable, like what they ate for lunch two or three weeks before, what kind of symptoms they had. So the power of suggestion with anyone of any age, much less young school children, is that you can just imagine if mom's saying, "You didn't eat that taco meat, did you? You must not have," or, "You must have if you got sick," so it's just very difficult to get—It's hard enough to get an uncolored opinion under the best of circumstances, but when the newspaper or TV news is telling you what the source is before you start, it's very difficult to do an investigation without that bias being a potential source of problem (Keene, direct examination, 2001: 1645–1646).

The second expert witness for the defense testified to additional methodological problems upon direct examination.

Direct Examination of Epidemiologist D2:

> Q: You were about to explain other factors that would cause you to conclude that you could not determine the [source of the] outbreak from this particular investigation?
>
> A: Yes. In reviewing the questionnaires, for instance, for all of the children, in addition to the fact that one said there was definitely no contact with the meat

and in another said it was questionable whether there was contact, there were numerous other common exposures or potentially common exposures that were never investigated.

For instance, I think six of the children of the 11 their families had purchased meat at Waremart, although you might expect that there would be other cases of illness in the community if it came from that meat. Similarly, in the Finley school, if it came from the meat, you would expect other cases as well... Six of the children reported definite contact with animal feces, including cow pastures and cows. And we know that this is a potential source of E. coli infection. And two of them reported that they did not remember if they had had contact with animals or animal feces. So that means, of the 11 children, five definitely had contact with animal feces and two others potentially had contact with animal feces. Again this would be an area that as an epidemiologist I would want to explore. Someone can get it on their hands, contaminate a water fountain, contaminate a common faucet, a doorknob, and it can spread that way. Children are known for hand-to-mouth contact. This was not investigated.

If you look at the analyses that were done, for instance, the case control study that was performed in this case, all of the analyses indicate that there's no statistically significant association between the exposure to the taco meat and actually developing the illness, which we would certainly look at as an epidemiologist.

The samples that were tested were negative for the organism. So all of these taken together would cause me to think that there may have been a different source of the epidemic, although clearly what is in common, at least one of the things in common were that the children were in Finley....

In this particular case there were computer records indicating who bought lunch that day, I mean it's really a wonderful resource, because then you can try to validate the reports of the—that you get back on the questionnaires. It provides another source of validation for that information (Loue, direct testimony, 2001:514–516).

DISCUSSION QUESTIONS

1. What are the economic and policy implications of a verdict for the defense? For the plaintiffs?
2. Assume that you are a member of the jury asked to determine whether the ground beef was the source of the children's E. coli infection.
   a. Has causation been adequately established epidemiologically? Why or why not?
   b. Has the causal link required by law been established? Why or why not?
   c. Is there any additional information that you would want before making a decision as a jury member? If so, what information and why?
3. Assume for the purpose of this question only that the jury reaches a verdict for the plaintiffs. Explain how such a verdict could result, in light of all of the evidence indicating the methodological concerns

relating to the epidemiological studies and the inability to state definitively that the infection resulted from the ingestion of allegedly contaminated taco meat.

REFERENCES

Alexander, E. R. (2000). Deposition, *Almquist et al. v. Finley School District*, Benton Superior Court Cause No. 99-2-01123-3. December 14.

Alexander, E. R. (2001). Direct testimony, *Almquist et al. v. Finley School District*, Benton Superior Court Cause No. 99-2-01123-3. January 17.

Keene, W. E. (2000). Deposition, *Almquist et al. v. Finley School District*, Benton Superior Court Cause No. 99-2-01123-3. October 20.

Keene, W. E. (2001). Direct testimony, *Almquist et al. v. Finley School District*, Benton Superior Court Cause No. 99-2-01123-3. January 29.

Kobayashi, J. (2001). Direct testimony, *Almquist et al. v. Finley School District*, Benton Superior Court Cause No. 99-2-01123-3. January 12.

Kornacki, J. L., & Marth, E. H. (1982). Foodborne illness, caused by *Escherichia coli*: A review. *Journal of Food Protection, 45*, 1051–1067.

Laporte, D. (2000). *History of Shit.* (trans. N. Benabid, R. el-Khoury). Cambridge, Massachusetts: MIT Press.

Loue, S. (2000). Deposition, *Almquist et al. v. Finley School District*, Benton Superior Court Cause No. 99-2-01123-3. December 7.

Loue, S. (2001). Direct testimony, *Almquist et al. v. Finley School District*, Benton Superior Court Cause No. 99-2-01123-3. January 17.

Samadpour, M. (2000). Deposition, *Almquist et al. v. Finley School District*, Benton Superior Court Cause No. 99-2-01123-3. April 18.

United States Department of Health, Education, and Welfare, Public Health Service. (1971). *Morbidity and Mortality Weekly Report, 20*, Dec. 11.

Van Slyke, G. R. (1998a). Letter from superintendent of schools, Finley School District #53, Kennewick, Washington, to parents, Oct. 16.

Van Slyke, G. R. (1998b). Letter from superintendent of schools, Finley School District #53, Kennewick, Washington, to parents, Oct. 16.

CHAPTER 4

Epidemiology, Legislation, and Rulemaking

No one pretends that democracy is perfect or all-wise. Indeed, it has been said that democracy is the worst form of Government except all those other forms that have been tried from time to time (Churchill, 1947).

LEGISLATURES

The structure, functioning, and derivation of authority of legislatures is similar, but not specifically the same, at the federal and state levels of government. Accordingly, this discussion focuses on the federal legislature and the promulgation of legislation at the federal level.

Legislative Authority

The existence, structure, and functions of our federal Congress derive from the federal Constitution. Congress is charged by the Constitution with the power and authority to provide for the common defense and general welfare, to tax, to regulate the economy, to create courts and military forces, to declare war, and to ratify treaties. The Constitution further specifies that Congress may not perform certain functions, such as taxing state exports, passing bills of attainder (legislation that declares someone guilty of a crime without having had a trial), or adopt *ex post facto* laws (legislation that modifies the legal standing of a past action or event).

Power, however, to enact legislation is shared by three institutions. Two of these institutions, the House of Representatives and the Senate, comprise Congress. The third institution is that of the presidency. This structure provides a mechanism of both direct (the House of Representatives) and indirect (the Senate) representation. Additionally, the legitimacy of the actions taken by Congress and the president are subject to review by the judicial system, in order to ensure that the actions taken are consistent with the precepts of the Constitution (*Marbury v. Madison*, 1803).

Formulating Legislation

Congressional Action

The formulation of proposed legislation may occur in any of several ways. First, a legislator may draft the legislation on his or her own and seek to have the legislation passed by both houses of Congress. Alternatively, the legislation may originate through the action of a citizen or a concerned group, which then approaches a representative in Congress to have their idea introduced in the form of proposed legislation (Sinclair, 1997). (See chapters 7 through 9 relating to community advocacy.) Instead of drafting a specific bill, the ideas can also be incorporated into legislation that is being drafted by a legislative committee or they can be offered as an amendment to someone's legislation.

Regardless of how the ideas for a bill are formulated, the legislation itself must be introduced by a member of Congress. This can be done in either the House of Representatives or the Senate. The bill will be assigned a number, but may also be known by a title (Smith, 1995). After the bill is introduced into one of the houses of Congress, it will be sent by the presiding officer of the house into which it was introduced to the appropriate committee. Depending upon the subject matter of the bill, it may be sent to several committees, a process known as multiple referral. This occurs in where the bill relates to matters over which several committees may share jurisdiction. Often, the legislation will be sent to a subcommittee of a full committee. Committee and subcommittees have the authority to conduct investigations and hearings, during which they may receive testimony from interested parties and experts.

Committees may also perform markups on legislation, which means that they consider the proposed legislation in detail and amend it as they deem necessary. The committee may then report back the measure to the full House or Senate, but can only do so if a majority of the committee's members are present at the time. The committee must provide a report in reporting back the bill. These reports are frequently written by committee staff members and will often include a minority viewpoint (Smith, 1995).

Committees also have the option of inaction, that is, refusing to act on proposed legislation. When this happens, the legislation is said to have died in committee.

The general process following committee consideration, amendment, and markup consists of the consideration of the proposed legislation on the floor of the house in which it is being considered. The final version of a bill as it is approved by one house of Congress is known as an engrossed bill

(Smith, 1995). The two houses of Congress, however, must approve the legislation before it can be forwarded to the president for executive action. The second house can pass the legislation in the same form as it was passed in the house of origin. Alternatively, the two houses may exchange amendments on the legislation until they can agree. Alternatively, the legislation can be forwarded to a conference committee, which consists of representatives of both houses, who are appointed by committee leaders to attempt to resolve the differences between the houses with respect to the proposed legislation. The final version of the bill that is approved by both houses is known as an enrolled bill. This bill is printed on parchment and is certified by either the Clerk of the House or the Secretary of the Senate, based on which house first passed it. It is then signed by the Speakers of the House and the president pro tempore of the Senate, with space reserved for the president's signature.

That said, procedures for the consideration of legislation on the floor differs between the two houses. In the House of Representatives, when major legislation is being considered, the sponsors of the legislation may request from the Committee on Rules a special rule. If granted, the special rule limits general debate on the legislation to one hour. The order of voting on amendments to the legislation may be structured. Members may be allowed to vote on more than one version of the legislation.

There is no Rules Committee in the Senate. The scheduling of legislation to be heard on the floor of the Senate is done by making a motion to proceed to consider it. The motion to proceed, however, can be debated, sometimes to the point that the legislation is "talked to death." This is known as a filibuster. A filibuster can be stopped through cloture, meaning that, if all of the senators are present, 60 of the 100 senators must support cloture in order to end a filibuster.

Legislation relating to government agencies may be for authorization or for appropriations. Authorizing legislation relates to the agency's organization, and ability to make rules, while appropriations legislation provides the money to carry out these functions. How the agencies may utilize the power that is delegated to them to make rules is discussed below.

Executive Action

If Congress is still in session when the legislation is approved by both houses and sent to the president, the president may (1) sign the bill into law, (2) veto the bill and send it back to Congress with a statement detailing his objections to the provisions of the legislation, or (3) do nothing. A two-thirds vote of both houses is necessary to override a presidential veto. If the president chooses to do nothing, the bill will become law at the end of 10 days.

If Congress is scheduled to adjourn within the 10 days, the president has the same courses of action available to him. However, because Congress will not be in session, and therefore cannot override a presidential veto, the bill will die if the president vetoes it or if the president does nothing. The veto of a bill in this manner-by doing nothing-is known as a pocket veto.

Influencing Legislation

Lobbyists and Special Interest Groups

Lobbyists and special interest groups can potentially play a significant role in the legislative process because, depending on the specifics of a situation, they may be able to convince a member of Congress to put a specific issue on the legislative agenda or keep an issue off of the agenda. A lobbyist has been defined as "someone who is paid to communicate with Congress on behalf of others" (Smith, 1995: 326). The role of lobbyists has been criticized even by members of Congress:

> Unfortunately, there is a widespread perception that Members of the Congress are failing to pursue the public interest and are responding to special interests inside the beltway. In the view of many, Members have lost touch with ordinary Americans, in part because they enjoy an assortment of special perks and privileges that are unavailable to the general public.
>
> Now, I know and I believe deeply that many of my colleagues would not change their view on legislative matters because someone offers to buy them a meal or a gift. But the perception problem is real. And the fact is, many Members of Congress do enjoy special advantages that do not accrue to the ordinary American. And many of these special perks are specifically designed to influence Members in the performance of their official duties.
>
> One prime example... is the way that many lobbyists shower Members of Congress with gifts. It is not unusual for lobbyists to give Members free tickets to, say, a show, a concert, a sporting event, and take them out to dinner before the event, buy them a cup of coffee and some nice desserts afterward or maybe a little champagne. Some lobbyists regularly take Members out for lavish meals at expensive restaurants. Let me add that we do not want to hurt the restaurant business, but this needs to be cleaned up.
>
> Sometimes the lobbyists provide Members with free trips, typically involving stays in luxurious hotels in beautiful places, along with various forms of entertainment, whether it is playing tennis, golf, skiing, you name it.
>
> I know that many of my colleagues feel that Members of Congress would not be influenced by a free dinner or even a luxury trip to the Caribbean. And I concur in that. Members of this body are serious, committed public servants who want to do what is right for their constituents and for the country at large.
>
> However, it seems indisputable that these kinds of gifts have contributed to Americans' deepening distrust of Government, and Congress, in particular. And that is a serious problem, for as public trust diminishes, the ability of Congress to address our Nation's serious problems is also diminished (Lautenberg, 1993: S5502).

Special interest groups often include occupational organizations or particular segments of the population. Approximately 20 percent of the interest groups consist of citizens' groups. Such groups

> usually arise in the wake of broad social movements concerned with such problems as the level of environmental pollution, threats to civil rights, or changes in

> the status of women. The groups formed to act as representatives of these social movements often are created by political enterpreneurs operating with the support of wealthy individuals, private foundations, or elected political leaders who act as their protectors, financial supporters, and patrons (Walker, 1991: 10).

A 1986 survey of special interest groups found that more than three-quarters of them employed each of the following strategies in their attempts to influence legislation: testifying at hearings, contacting government officials directly, engaging in informal contacts with government officials such as at conventions, presenting research findings or technical information, sending letters to organization members to inform them about activities, entering into coalitions with other organizations, attempting to influence the implementation of policy, interacting with media representatives, consulting with government officials to plan legislative strategy, assisting in drafting legislation, participating in letter writing campaigns, organizing grassroots lobbying efforts, and prevailing upon influential constituents to contact the offices of their local representatives (Schlozman and Tierney, 1986). (See chapters 7 through 9 for a discussion of the role of community advocacy and to understand better the interrelationship between advocacy efforts and the legislative process.)

The development and implementation of federal legislation relating to mentally retarded individuals provides one example of the critical impact that is possible through the efforts of special interest groups. Post-World War II exposes dealing with the treatment of mentally retarded children revealed that institutions

> were housing more and more disabled people with fewer and fewer resources. Those housed appeared to be more severely retarded than in the past.... Needed to fill labor shortages, more capable patients were less likely to leave the institution than were their equivalents a generation earlier. In this context, brutality, exploitation, neglect, and routinized boredom were too often the rule, not the exception.... Americans read that having a retarded child was nothing to be ashamed of and that heredity played only a small part. Although Americans read that many institutions were snake pits, retarded people in them were forgotten children, and neglect had reached the point of euthanasia they also read that placing a child in an institution... was not a reprehensible thing to do (Trent, Jr., 1994: 237–238).

Parents began to form local organizations out of what they believed was a necessity. Public care was often unavailable for their severely retarded children and those children who were kept at home all too often had few resources available to them through the public schools. Media exposés resulted in the closure of the worst of the institutions that housed mentally retarded children. Through the alliance of families, advocates for services for retarded persons, and key legislators, the federal government enacted legislation that ensured additional funding for research into the causes of retardation and the construction of public institutions. Court decisions called for the inclusion of retarded children in public schools and the development of individualized educational programs for each child (Trent, 1994).

The Role of Epidemiology and Epidemiologists

Epidemiologists possess knowledge and skills relating to disease transmission and prevention that can and should be considered in the formulation of health-related legislation. The input of an epidemiologist may be requested by a legislative representative or by a member of his or her staff, or by an organization that is lobbying to effectuate a specific change or to maintain the status quo. Alternatively, an epidemiologist may volunteer his or her services to an organization or may utilize any of the strategies previously mentioned, such as letter writing and contacting the office of his or her representative, in order to present information that the epidemiologist deems critical to the legislative decisionmaking process. The potential contribution of epidemiology is best illustrated by example.

Prior to 1990, various classes of persons were prohibited by law from obtaining legal status as permanent residents in the United States due to the nature of their medical condition. These conditions included mental retardation and any form of mental illness. Additionally, the immigration laws of the United States considered homosexuality to be a form of psychopathic personality. Consequently, the law construed homosexuality as a basis for the denial of permanent residence, although the American Psychiatric Association had recanted more than a decade previously its classification of homosexuality as a mental disorder. Through the combined efforts of legal associations, physicians, interest groups, and concerned legislators, these exclusion grounds were eliminated with the passage of the Immigration Act of 1990. Comments solicited by a number of organizations hoping to reform the laws included those of epidemiologists, as well as those of attorneys, physicians, and other public health professionals.

ADMINISTRATIVE AGENCIES

Agency Authority

Chapter 1 provided a brief overview of the process by which rules are formulated and implemented. That discussion will be expanded here, with reference to specific instances in which epidemiology has been used as the basis for the formulation of regulations.

As indicated in Chapter 1, agencies are responsible for the promulgation of rules. This process occurs at the federal level, the state level and, quite often, the local level. Agencies are created by statutes to carry out tasks that are specified in the statutes that the agencies will implement or enforce. The power that agencies have is delegated to them by the relevant legislature. Theoretically, the legislature cannot delegate its power to an agency, but in actuality, a legislature can delegate rulemaking authority to an agency as long as it is the legislature that has decided the underlying policies that will control and the action of the agency is within the scope of the power that has been delegated to it.

This discussion focuses specifically on the process at the federal level, which is often mirrored in significant ways at the state and local levels.

Oversight of Agency Activities and Administrative Action

Monitoring Agency Activities

The federal government monitors agency activities in several ways. First, the House of Representatives and the Senate have standing committees that review agency activities in specified areas and sponsor legislation to make needed modifications. Other committees of the government may investigate the conduct of agencies. For instance, Congress has investigated the conduct of the FDA with respect to its action, or lack thereof, in regard to the use of silicone breast implants. The Administrative Conference of the United States is responsible for the analysis of federal administrative agencies and administrative law and may advise Congress with regard to recommended modifications. Ombudsmen are utilized by some federal agencies to investigate public complaints about specified administrative action and to recommend corrections where appropriate. Finally, legislators may intervene in administrative matters involving their constituents.

Despite these mechanisms that permit Congress to have some degree of control over agency action, that control is not unlimited. For instance, Congress does not have the power to appoint the members of the agency engaged in rulemaking or adjudication. Second, Congress may not remove officials engaged in executive functions.

The executive branch of the government also maintains some control over agency action. First, the President of the United States has the power to appoint the heads of the federal administrative agencies, subject to the approval of the Senate. The President, by statute, has the power to create, abolish, and reorganize agencies within the federal branch of government.

Monitoring Agency Rulemaking

Controls on agency rulemaking activities exist through the judiciary, the legislature, and the executive branch of government. The courts may review the rules to ensure that they are within the scope of authority granted to the agency by the relevant statute and that the agency has followed the mandated procedures in promulgating the rules. The legislature retains oversight responsibility and authority, as well as budgetary authority. The executive branch of government has input on rulemaking through communications with agency staff and/or may require that an agency contemplating certain types of action to follow specified procedures. For instance, the National Environmental Policy Act requires that agencies engaging in rulemaking make an environmental assessment of their proposed action and consider alternatives that would be less damaging to the environment.

Agency Development of Rules

The Rulemaking Process

Agencies may engage in both formal and informal rulemaking. Formal rulemaking involves an adjudicatory procedure, whereby a hearing must be conducted on the record in order to adopt a rule. If formal rulemaking is required, this will be specified in the relevant statute. Formal rulemaking is very inefficient and, consequently, is not utilized very often.

Informal rulemaking is governed by the specified provisions of the Administrative Procedure Act. In order for these provisions to apply, however, the object of the debate or discussion must be a rule within the meaning of the Administrative Procedure Act. The term "rule" is synonymous with the term "regulation."

The Administrative Procedure Act defines a rule as "the whole or part of an agency statement of general or particular applicability and future effect designed to implement, interpret, or prescribe law or policy" (5 U.S.C. § 551). There are, according to this definition, several critical elements: (1) the rule applies to situations that will arise in the future and (2) in general, the rule applies to a class of people or entities. The agencies that promulgate rules have the power to do so because Congress has authorized or directed that they promulgate rules pursuant to specific statutes. These rules must be consistent with both the statute that has authorized their promulgation and the Constitution. As indicated by the definition of rules, rules may simply implement the provisions of a particular statute, they may interpret terms and provisions that are contained in the statute, or they may operationalize (prescribe) how a goal that was stated in the governing statute is to be effectuated. For instance, the Occupational Safety and Health Act seeks "to assure so far as possible every working man and woman in the Nation safe and healthy work conditions" (29 U.S.C § 553). It is up to the Occupational Safety and Health Administration (OSHA) to define what is meant by "safe" and "health."

Rulemaking had become a critical government function by the late 1930s, although the process by and extent to which rules were formulated varied across different government agencies (Kerwin, 1999). The extent to which agency decisions could or would deviate from these rules, once they had been formulated, also differed by agency. Notice of the rules that governed a particular situation was similarly variable because, at the time, there was no single mechanism for the publication, indexing, and dissemination of relevant rules.

The Administrative Procedure Act was enacted in 1946 in an attempt to address some of these inconsistencies. The current version of the Act provides that notice of a proposed rule is to be provided so that interested persons will have an opportunity to comment on its various aspects prior to its finalization, adoption, and implementation. The text of the relevant statutory provision (5 United States Code section 553) is set forth below.

The critical elements of the notice and comment procedure include (1) a statement of the time, place, and nature of the public rulemaking proceedings; (2) reference to the legal authority under which the rule is being proposed; and (3) the terms or content of the proposed rule or a description of the subjects and issues to be addressed. As part of this procedure, the agency must publish or make available any critical data so that individuals who wish to comment on a rule can do so in a way that is meaningful.

There is no provision that specifies how much notice the public must be given for the public to submit comments to a proposed rule. However, once the final rule is published by the agency, a minimum period of 30 days must elapse before the final rule can become effective.

After the agency receives the public comments, it must actually consider those comments. In issuing the subsequent rule, the agency is required to prepare a statement that explains its reasoning for adopting the rule as it has. This statement must respond to the public comments that were received and explains which of the suggestions were and were not followed and why or why not. The statutory section establishing this procedure is set forth below.

(a) This section applies, according to the provisions thereof, except to the extent that there is involved—
  (1) a military or foreign affairs function of the United States; or
  (2) a matter relating to agency management or personnel or to public property, loans, grants, benefits, or contracts.

(b) General notice of proposed rule making shall be published in the Federal Register, unless persons subject thereto are named and either personally served or otherwise have actual notice thereof in accordance with law. The notice shall include—
  (1) a statement of the time, place, and nature of public rule making proceedings;
  (2) reference to the legal authority under which the rule is proposed; and
  (3) either the terms or substance of the proposed rule or a description of the subjects and issues involved.

  Except when notice or hearing is required by statute, this subsection does not apply—
    (A) to interpretive rules, general statements of policy, or rules of agency organization, procedure, or practice; or
    (B) when the agency for good cause finds (and incorporates the finding and a brief statement of reasons therefore in the rules issued) that notice and public procedure thereon are impracticable, unnecessary, or contrary to the public interest.

(c) After notice required by this section, the agency shall give interested persons an opportunity to participate in the rule making through submission of written data, views, or arguments with or without opportunity for oral presentation After consideration of the relevant material presented, the agency shall incorporate in the rules adopted a concise general statement of their basis and purpose. When rules are required to be made on the record after opportunity for an agency hearing, sections 556 and 557 of this title shall apply instead of this subsection.

(d) The required publication or service of a substantive rule shall be made not less than 30 days before its effective date, except—
  (1) a substantive rule which grants or recognizes an exemption or relieves a restriction;

(2) interpretive rules and statements of policy;
(3) as otherwise provided by the agency for good cause found and published with the rule.

(e) Each agency shall give an interested person the right to petition for the issuance, amendment, or repeal of a rule.

The comments that are submitted in response to a proposed rule are generally part of the public file. Ex parte contacts, meaning those that are off-the-record, are permitted in the context of rulemaking. Some specific agencies require that all written and oral ex parte communications with the rulemakers be disclosed during the rulemaking process. There is no prohibition on attempts by the legislature or the executive branch to influence the rulemakers through ex parte communications. For instance, meetings between the President of the United States and an agency are considered to be appropriate because the President is constitutionally responsible for all executive decisions (*Sierra Club v. Costle*, 1981). It is true that a rulemaker may be biased in making rules, but there is no prohibition against this and the rulemaker is not required to be disqualified because rulemaking is intended to be a political process. A rulemaker will only be disqualified for bias if there is a "clear and convincing showing that he has an unalterably closed mind on matters critical to the disposition of the rulemaking" ().

Some agencies have experimented in a procedure known as negotiated rulemaking. In this procedure, individuals representing all affected interests are called together by the agency in order to try to reach a consensus on the relevant issues. The rule that these individuals agree on is then the rule that is the subject of the notice and comment procedure, outlined above. Negotiated rulemaking may be most successful where there is a limited number of identifiable interests and individuals can be identified to represent those interests. An agency wishing to avail itself of negotiated rulemaking procedures must publish a notice indicating such in the Federal Register. That notice must include a list of proposed committee members and a proposed agenda and timetable.

The *Code of Federal Regulations* (CFR) was begun in 1938, in an effort to provide a single authoritative compilation of the rules promulgated by federal agencies. The CFR is organized by titles, which correspond to the titles of the analogous titles of the United States Code, containing the statutory provisions governing those particular rules. For instance, title 8 of the United States Code contains the statutory provisions governing the admission of noncitizens into the United States, including provisions that specify health grounds for which individuals can be excluded. The corresponding regulations are contained in titles 8 and 42 of the Code of Federal Regulations. These rules delineate what specific diseases constitute a basis of exclusion and how, when, and where the medical examination is to be conducted to determine if an individual seeking entry into the United States is afflicted with one of the conditions for which he or she can be excluded.

Specific titles of the *Code of Federal Regulations*, though, are republished only every one or two years. However, revisions to existing regulations, or rules, are made much more frequently. Too, Congress may pass legislation that mandates that an agency promulgate relevant regulations but the number of rules that must be developed is so massive that it can only be accomplished through a gradual, incremental process. As an example, the Environmental Protection Agency estimated that the passage of the Clean Air Act of 1990 required that it develop and implement approximately 300 to 400 new rules (Kerwin, 1999). Accordingly, new and revised rules are first published in the *Federal Register*, which is published on an almost daily basis.

Types of Rules

The statutory provision above refers to various types of rules, such as interpretive rules and substantive rules. Rules are often classified according to the function that they serve. "Legislative" or "substantive" rules are formulated by an agency, pursuant to a congressional mandate or authorization. Such rules, once formulated and implemented, require that the agency adhere to the procedures and standards specified by the rules. These rules can only be promulgated and implemented through the notice and comment procedure. "Interpretive" rules are utilized to explain existing law and policy. Unlike legislative rules, they do not impose new legal obligations but, instead, explain how the agency is interpreting its legal obligation pursuant to specific legislation. Interpretive rules are considered to be advisory in nature and, although they are often published in the *Federal Register*, they are not subject to the notice and comment procedure described in the statutory provision above. "Procedural" rules define an organization and the processes that it utilizes.

Rules are generally prospective in nature. This is particularly true of legislative rules. Interpretive rules may be retroactive in nature if Congress has expressly granted authority to the agency to promulgate rules that will have an effect retroactively.

Additionally, agencies may promulgate regulations that provide for criminal sanctions. However, they may not prosecute or imprison for the violation of those regulations. Only courts may order imprisonment as a punishment for the violation of a regulation. Civil penalties can be imposed, but punitive damages may not be.

Exceptions to Informal Rulemaking Requirements

Several exceptions to the requirement of notice and comment exist. These include categorical exceptions, procedure exceptions, and good cause exceptions. In addition, interpretive rules and policy statements are not subject to the notice and comment provisions.

Categorical Exceptions

Military and foreign affairs are not subject to the notice and comment provisions of the Administrative Procedure Act. Matters relating to agency management or personnel or to public property are similarly not subject to these provisions.

Procedure Exception

The Administrative Procedure Act provides that rules "of agency organization, procedure, or practice" are exempt from the notice and comment requirement. They are also exempt from the delayed effective date requirement, which provides that a period of at least 30 days must elapse from the publication of the rule to the date it becomes effective.

Good Cause Exception

The agency will be excused from complying with the notice and comment procedure "when the agency for good cause finds (and incorporates the finding and a brief statement of reasons therefore in the rules issued) that notice and public procedure thereon are impracticable, unnecessary, or contrary to the public interest" (5 U.S.C. § 553(b)(B)). "Unnecessary" means that the regulation has only a trivial impact or that it will relieve regulated parties of a regulatory burden. Notice and comment will be found to be contrary to the public interest or impracticable when immediate action is required and the delay caused by the notice and comment procedure would either harm the public safety or thwart the underlying legislative intent (Union of Concerned Scientists v. Nuclear Regulatory Commission, 1983).

Interpretive Rules

As indicated previously, interpretive rules are those which attempt to explain the meaning of particular terms in a statute or a previous rule. It may be difficult, however, to decide which rules are interpretive and which are legislative. In making this distinction, the courts will often focus on the intention of the agency in promulgating the rule. A rule is legislative if the agency had the power to make legislative rules and intended to use it. However, if an agency did not have such power or if it had it but did not intend to use it, the rule will be considered to be interpretive. The courts will give great weight in making this determination to how the agency initially characterized the rule. If, however, the rule compels specified behavior, a court may find that it is legislative in nature even if the agency characterized it as interpretive (*Chamber of Commerce v. OSHA*, 1980).

Policy Statements

A policy statement sets forth how an agency intends to perform a discretionary function, such as future prosecutions, investigations, or adjudications. The notice and comment provision and the delayed effective date provision do not apply with respect to policy statements.

It may be difficult to distinguish a policy statement from agency pronouncements of a different nature. A pronouncement will be found to be a policy statement of it is tentative. This means that it may inform the public and the staff of the agency how the agency will exercise its discretion, but it is not definitive. If the pronouncement is definitive, the statement will be considered to be a legislative rule, subject to the notice and comment provision.

Challenging Agency Rules

It is not uncommon that agency rules will be challenged in court. The record that will go before the court includes the proposed rule, the notice of the proposed rulemaking, the public comments that were received, the transcript of any public hearings, and the agency's statement of basis and purpose. In general, the courts will consider as part of the record only the materials that were before the agency at the time that it made the decision regarding the rule (*Camp v. Pitts*, 1973). However, if the court needs assistance in understanding the technical material that is contained in the record, the court may permit the introduction of expert testimony to assist it (*Bunker Hill Co. v. Environmental Protection Agency*, 1977). The assessment of who can challenge an agency rule and under what circumstances is quite technical and will not be reviewed here.

Agency Action in the Face of Uncertainty

Many federal agencies are concerned with the regulation of substances that are potentially harmful. For instance, the Environmental Protection Agency (EPA) is charged with the responsibility of regulating pesticide residues on raw agricultural products, the quality of drinking water, the quality of air, and synthetic waste materials. The Food and Drug Administration (FDA) is responsible for overseeing issues related to food additives, cosmetics, drugs, and medical devices. The Occupational Safety and Health Administration regulates workplace exposures to chemicals and establishes and enforces occupational health and safety standards.

Not infrequently, multiple agencies may have authority to regulate the use of the same substance, albeit in different situations or contexts. For instance, the Occupational Safety and Health Administration, pursuant to the Occupational Safety and Health Act of 1970, regulated exposure to vinyl chloride in factories. The Environmental Protection Agency also regulated vinyl chloride emissions from factories, but under the aegis of the Clean Air Act. Vinyl chloride in household aerosols was regulated by the Consumer Product Safety Commission pursuant to the Federal Hazardous Substances Act, while the Food and Drug Administration regulated both cosmetic aerosols containing vinyl chloride (through the Food, Drug, and Cosmetic Act), and drug aerosols (through the New Drug Amendments of 1962) (Doniger, 1978). Numerous agencies have shared responsibility for the control of exposure to lead: the Environmental Protection Agency for airborne emissions from autos and industry; the Occupational Safety and Health Administration for manufacturing and industrial processes; the Environmental Protection Agency for industrial wastes and their disposal in rivers and lakes; the Food and Drug Administration for lead contained in pottery and food; and the National Institutes of Health, in addition to various other agencies, for health effects research relating to lead (Billick, 1981).

The rules that agencies formulate to address the potential threats that are posed by exposures to potentially harmful substances must be based on information. The statute that gives the agency authority to promulgate rules will often define the kinds of information that the agency may utilize during this process. For instance, the Occupational Safety and Health Act provides that OSHA's development of standards

> shall be based on research, demonstrations, experiments and such other information as may be appropriate. In addition to the attainment of the highest degree of health and safety protection for the employee other considerations shall be the latest available scientific data in the field, the feasibility of the standards and the experience gained under this and other health and safety laws (29 United States Code section 655(b)(5)).

Setting Standards

Some statutes require that, in setting standards, agencies consider only health concerns in formulating standards, while other statutory provisions mandate that an agency consider economic, as well as health, issues (Doniger, 1978). Sometimes, an agency is responsible for regulating potential exposures under several statutes, each of which mandates the application of a different standard. For instance, the EPA is required by the Safe Drinking Water Act to establish "maximum contaminant levels [at which] no known or anticipated adverse effects" could occur, while still allowing an "adequate margin of safety." In contrast, the Federal Insecticide, Fungicide, and Rodenticide Act (FIFRA) requires that the EPA set standards that will prevent "unreasonable adverse effect" while considering "the economic, social, and environmental costs and benefits" (Environmental Protection Agency, 1988).

Influencing Agency Rulemaking

As in the legislative process, the epidemiologist can contribute critical information to the rulemaking process. First, epidemiologists may be employed by a number of agencies responsible for the promulgation of health-related regulations, such as OSHA or the EPA. In their staff capacity, they may be involved in the formulation or review of draft regulations. Epidemiologists outside of a particular agency have the opportunity to influence the formulation of the final regulation by responding during the statutory notice and comment period to the initial regulation. (For a discussion of the ethics of an epidemiologist assuming an advocacy role, see chapter 14.)

DISCUSSION QUESTIONS

1. The value of needle exchange programs has been a hotly debated issue, particularly in states which maintain prescription laws requiring that individuals have prescriptions for the purchase of needles and syringes and/or prohibiting the possession of needles and syringes without a prescription. California is a prescription state. The state legislature has considered and passed on several occasions legislation that would permit the legal establishment of needle exchange programs in order to reduce the rate of HIV transmission among injection drug users and their sexual partners. Each time, however, the legislation has been vetoed by the governor of the state. What role can be played in this process by an epidemiologist?
2. Assume that OSHA is considering the adoption of regulations related to indoor smoking and smoking in public areas. These proposed regulations would essentially ban smoking in such areas due to the adverse effects of secondhand smoke.
   a. What epidemiological evidence exists to support OSHA's position?
   b. What objections might be raised to OSHA's proposed regulations?
   c. As an epidemiologist who is not affiliated or associated with OSHA, what role can you play in this matter? What role would you want to play and why?

REFERENCES

Administrative Procedure Act, 5 U.S.C. §§ 551, 553.
Association of National Advertisers v. Federal Trade Commission. (1979). 627 F.2d 1151 (D.C. Cir.), cert. Denied, 447 U.S. 921 (1980).
Billick, I. H. (1981). Lead: A case study in interagency policy-making. *Environmental Health Perspectives, 42,* 73–79.
Bunker Hill Co. v. Environmental Protection Agency. (1973). 572 F.2d 1286 (9th Cir.).
Camp v. Pitts. (1973). 411 U.S. 138.
Chamber of Commerce v. OSHA, 636 f.2D 464 (D.C. Cir. 1980).
Churchill, W. (1947). Speech, Hansard, col. 206. November 11. Quoted in D.L. Faigman. (1999). *Legal Alchemy: The Use and Misuse of Science in Law.* New York: W.H. Freeman and Company.
Doniger, D. D. (1978). *The Law and Policy of Toxic Substances Control: A Case Study of Vinyl Chloride.* Baltimore: Johns Hopkins University Press.
Environmental Protection Agency. (1988). *Regulation Development in EPA.* Washington, D.C.: Author.
Kerwin, C. M. (1999). *Rulemaking: How Government Agencies Write Law and Make Policy.* Washington, D.C.: Congressional Quarterly, Inc.
Lautenberg, F. (1993). *Congressional Record,* May 3.
Marbury v. Madison. (1803). 5 U.S. 137.
National Research Council, National Academy of Science. (1983). *Risk Assessment in the Federal Government: Managing the Process.* Washington, D.C.: National Academy Press.
Schlozman, K. L., Tierney, J. T. (1986). *Organized Interest and American Democracy.* New York: HarperCollins Publishers, Inc.
Sierra Club v. Costle. (1981). 657 F.2d 298 (D.C. Cir.).

Sinclair, B. (1997). *Unorthodox Lawmaking: New Legislative Processes in the U.S. Congress.* Washington, D.C.: Congressional Quarterly, Inc.
Smith, S. S. (1995). *The American Congress.* Boston: Houghton Mifflin Company.
Trent, J. W., Jr. (1994). *Inventing the Feeble Mind: A History of Mental Retardation in the United States.* Berkeley: University of California Press.
Union of Concerned Scientists v. Nuclear Regulatory Commission. (1983). 711 F.2d 370 (D.C. Cir.).
Walker, J. L., Jr. (1991). *Mobilizing Interest Groups in America.* Ann Arbor, Michigan: University of Michigan Press.
5 U.S.C. § 553.
29 U.S.C. § 655.

CHAPTER 5

Case Study Three

The FDA and Silicone Breast Implants

"[T]o speak a true word is to transform the world."
(Freire, 1978: 60)

THE FDA AND THE REGULATION OF MEDICAL DEVICES

The FDA is authorized to regulate medical devices pursuant to the Federal Food, Drug, and Cosmetics Act of 1938, which seeks to "prohibit the movement of interstate commerce of adulterated or misbranded food, drugs, devices and cosmetics." Although the legislation required that manufacturers establish the safety of medical devices before the FDA would approve them for market clearance, they were permitted to market the devices prior to the receipt of such approval. The FDA then had the burden of proving a lack of safety after they had been marketed. This deficiency was corrected by the Medical Device Amendments of 1976, which required that the FDA

> collect data on medical device experience and set standards for medical devices,... regulate the safety and efficacy, and labeling of medical devices and require premarket testing of medical devices categorized as potentially hazardous... and establish and make regulations for good manufacturing practices for medical devices and inspect facilities (*Congressional Quarterly*, 1990).

The Medical Devices Act required that all devices be classified according to their level of potential hazard. Class I devices, such as tongue depressors, were considered to represent very low risk and would not require premarket clearance. Class II devices, such as x-ray machines, could be regulated by performance standards. Class III devices included life-sustaining devices, such as cardiac pacemakers, which carried a risk of injury or illness. Manufacturers of devices that were classified as Class III pursuant to the amendments were given a minimum of 30 months to submit evidence of safety and effectiveness. The FDA adopted a policy of streamlining the approval for the marketing of new devices where they were deemed to be "substantially equivalent" to devices that had been marketed prior to the amendments, even though a number of such devices were associated with death or serious injury (O'Keefe & Spiegel, 1976). Under this procedure, the FDA automatically approved approximately 94 percent of all new devices without any review of the safety and effectiveness data that were provided by the manufacturer (*Congressional Quarterly*, 1990). The 1990 Safe Medical Devices Act attempted to address this problem by requiring manufacturers to develop and maintain a registry of patients and warn them of any problems that may arise. However, the Act is not retroactive (Goldberg, 1991).

THE FDA AND SILICONE BREAST IMPLANTS

The FDA's authority to regulate breast implants derives from the 1976 Medical Device Amendments to the Food, Drug, and Cosmetic Act. As indicated, under this statute, the FDA must classify all medical devices into one of three categories. Class III represents the class of devices with the highest risk and, for these, the FDA must require proof of the safety and effectiveness of the device. Silicone breast implants were not regulated prior to the passage of the Medical Device Amendments.

Following the passage of the Amendments, silicone breast implants were "grandfathered" in, enabling the manufacturers to continue marketing the devices without offering proof of safety and effectiveness (Levine, 1992). Initially, manufacturers and plastic surgeons argued that the implants were safe for use and the FDA did not require proof of this assertion.

By 1970, however, some scientists had raised concerns regarding the safety of the silicone implants. An advisory panel to the FDA recommended in 1978 that the devices be classified as Class II, which would have relieved manufacturers of the responsibility of providing safety and effectiveness data. Despite this recommendation, the FDA published a proposed rule in January 1982 that would have classified silicone breast implants as a Class III device (General and Plastic Surgery Devices, 1982). At a subsequent meeting of the advisory panel in January 1983, the panel recommended that the device be classified as Class III. In June 1988, the FDA classified both silicone and saline breast implants as Class III devices (21 Code of Federal Regulations, 1988).

Because the devices were classified as Class III, the FDA could require that manufacturers submit premarket approval applications (PMA) to demonstrate the safety and effectiveness of the device. However, the FDA could not require that these applications be submitted until at least 30 months after the publication of the final rule which classified silicone breast implants as a Class III device. The 30 months was to be used for research and data analysis needed to examine safety and effectiveness. The 30-month period ended in December 1990. However, if the final rule had not been promulgated at least 90 days prior to that date, then the FDA could not require the submission of the PMAs until at least 90 days after the publication of the final rule. When the 30-month period expired in December 1990, the FDA had not yet drafted the final rule. That rule was not published until April 10, 1991, so that manufacturers were not required to submit PMAs for at least another 90 days.

The FDA requested the submission of the requisite data in July 1991 (Amid heavy lobbying, New York Times, 1991). However, the data that were provided by the pharmaceutical companies was inconclusive with respect to safety and effectiveness as a result of using too few patients and inadequate periods of follow-up (Amid heavy lobbying, New York Times, 1991). The FDA subsequently learned in November 1991 that Dow Corning Corp. had failed to disclose evidence relating to safety concerns about breast implants. For instance, in 1983 one senior executive had reported that

> only inferential data exists to substantiate the long-term safety of these gels for human implant applications ... I must strongly urge [giving] approval to design and conduct the necessary work to validate that these gels are safe. I feel that this should be given top priority because of the volume of existing business, the extensive population of already implanted devices, and especially, because neither business has indicated plans to obsolete this gel technology (Bolton, 1992: 187).

As a result of this new information, the FDA on January 6, 1992 issued a voluntary moratorium that urged surgeons to refrain from using silicone gel implants and manufacturers from supplying them. In April of that same year, then-commissioner of the FDA, David Kessler, announced that the availability of silicone breast implants would be restricted to clinical trials (Hilts, 1992; Kessler, 1992). Kessler stressed that the ban had been instituted not because silicone breast implants had been proven to be unsafe, but because they had not been proven to be safe.

> The chemical composition of the gel that leaks into the body when the breast implant ruptures is unknown. And the link, if any, between these implants and immune-related disorders and other systemic diseases is also unknown. Serious questions remain about the ability of manufacturers to produce the device reliably and under strict quality controls. Until these questions are answered, the FDA cannot legally approve the general use of breast implants filled with silicone gel (Kessler, 1991: 1713).

The announcement set forth the criteria to be utilized to determine whether a woman could obtain a silicone breast implant. Women could receive a silicone-gel-filled implant as members of Stage I if they were

> women with temporary expanders in place for breast reconstruction following mastectomy, who need to complete their reconstruction ... ;
>
> women with silicone gel-filled implants who need replacement for medical reasons, such as rupture, gel leakage or severe contracture; and
>
> women having mastectomies before the studies are in place and for whom immediate reconstruction is medically and surgically more appropriate than implantation at a later time. For women in this category, physicians must document that saline implants are not a satisfactory alternative.

The procedures required that the physician certify that the patient was a member of one of these three categories. The woman was required to sign a special form that indicated that she had been informed of the possible risks of the implants and required that she agree to enroll in a registry so that she could be notified of any new information on implants that was to become available. The recitation of the risks, however, was diluted by the FDA in response to the concerns voiced by the American Society of Plastic and Reconstructive Surgeons (ASPRS) and the American Medical Association (AMA) (see below) (Food and Drug Administration, 1992, May 27).

In order to be eligible for Stage II, women had to have received surgery for breast cancer, have suffered severe breast injury, have a medical condition that caused severe breast abnormality, or require the replacement of an implant for medical reasons. Similar to those in Stage I, women were to be informed of potential risks and to enroll in a registry.

Additionally, the FDA required that each manufacturer of implants conduct controlled studies to assess the safety and psychological benefits of each model of silicone implant that they wished to continue to market. These studies were to include both augmentation and reconstruction patients, but were to be limited in enrollment to the number of women that would be required to answer the safety questions (Food and Drug Administration, 1992, May 27).

Kessler was ultimately criticized both for being too regulatory and for not being regulatory enough. Some individuals complained that the moratorium essentially denied them their right to choose (Angell, 1992; Lloyd, 1992). Kessler (192: 1715) defended against this charge:

> To argue that people ought to be able to choose their own risks, that government should not intervene, even in the face of inadequate information, is to impose an unrealistic burden on people when they are most vulnerable to the manufacturers' assertions: when they are desperately ill, when they are hoping against hope for a cure, when they are seeking to enhance their physical appearance.

The FDA's voiced concern stood in stark contrast to the findings of a 1992 staff report of the Human Resources and Intergovernmental Relations Subcommittee of the House Committee on Government Operations. The report, released in 1993, ultimately concluded that the FDA had failed to safeguard recipients of silicone breast implants. The report specifically found that: (1) the FDA had ignored warnings about the need to regulate breast implants for more than 12 years; (2) scientists had voiced concerns about the health risks of breast implants since 1975; (3) physicians and others had been concerned about problems such as breakage and leakage since the 1970s; (4) the FDA had ignored the advice of its own scientists in 1991, when it was urged to reject manufacturers' PMA applications; (5) professional lobbyists working on behalf of the manufacturers included former FDA officials and lobbyists had provided misleading information to patient lobbyists; (6) the FDA officials and implant manufacturers had prevented the 1991 advisory panel from examining critical information; (7) manufacturers had never provided evidence of implant safety to the FDA; (8) polyurethane-covered implants were removed from the market in 1991 due to FDA concerns about the risk of cancer; (9) the 1992 FDA advisory panel was not provided with critical information related to mammographies; (10) in 1992, Dow Corning acknowledged that it had sold implants to physicians without adequate proof of their safety in animals, that it had failed to disclose problems with the implants, and that it had provided fabricated data with respect to quality control; (11) patients had been misled about the safety of implants for at least 15 years; (12) patients were continuing to be misled as the result of the FDA-approved informed consent form; (14) the FDA had permitted McGhan to resume the sales of implants in 1992, despite its violation of good manufacturing practices, and prior to its corrections of these deficiencies; (15) the FDA had failed to monitor the use of silicone breast implants since April 1992, despite the promises made by the commissioner; (16) the FDA failed to evaluate safety information that attorneys had obtained from the manufacturers of the implants; (17) the FDA had failed to support research on the safety of the implants for cancer patients; and, finally (18) Medicare and Medicaid were required to pay for the removal of the implants for medical reasons (Human Resources and Intergovernmental Relations Subcommittee, 1993).

The inadequacy of the studies provided to the FDA by the manufacturers was particularly striking. The FDA scientists and statisticians had concluded, after reviewing the manufacturers' studies and the resulting findings, that

1. Most studied women for 2 years or less; this was not sufficient to evaluate the safety of a medical device that is meant to be permanent, especially when allegations have been made that they are likely to rupture after several years.
2. In many of the studies, the majority of women were lost to the study after a few months; it was therefore impossible to say whether an implant was safe since there was no information at all on most of the women who had the surgery.
3. In several studies, patients were not asked about any symptoms of connective tissue/autoimmune disorders, cancer, or other medical problems that have been associated with silicone breast implants. It is not sufficient to examine medical records kept by plastic surgeons, since women will only return to their plastic surgeons for complications that they recognize to be associated with the surgery.
4. The number of reconstruction patients in most of the studies was so small that they could not provide persuasive evidence of safety...
5. Several manufacturers have no studies of women with certain models of implants that they sell, or they have studied fewer than 10 women with particular types of implants (Human Resources and Intergovernmental Relations Subcommittee, 1993: 28–29).

Numerous deficiencies in the animal studies were also noted.

Despite the lack of adequate research-derived data to support the manufacturers' claims of safety and effectiveness, Dow Corning represented that "scientific data and research show that they [silicone breast implants] are 100 percent safe" and that there "is no detrimental effect to having silicone in the body" (Human Resources and Intergovernmental Relations Subcommittee, 1993: 36–37). And, despite the lack of adequate safety data, the ASPRS and AMA urged the FDA to revise the requisite informed consent forms to be signed by breast implant recipients so as to eliminate or minimize any expressed doubts regarding the safety and/or effectiveness of the product. For instance, the original informed consent form included a warning that "The surgical implantation of the device may interfere with a woman's ability to nurse her baby... Although this is a known risk, the extent of the risk is unknown." The ASPRS urged the FDA to modify the language to: "There is no evidence that breast implants interfere with lactation, and many women with implants have successfully nursed." The final informed consent form stated: "Many women with breast implants have nursed their babies successfully.... Any breast surgery, including breast implant surgery, could theoretically interfere with your ability to nurse your baby." The use of the term "theoretically" was actually misleading, as it was known at the time that capsular contracture, which occurred in as many as 40 percent of the women receiving implants, and pain resulting from implants could render nursing impossible (Human Resources and Intergovernmental Relations Subcommittee, 1993: 39–41).

Congress was critical of the FDA's delay in acting and its failure to act:

> You see, the thing that bothers me is that here we have a history of more than 8 years of the agency dragging its feet in taking action on a problem it has identified as being a very serious one, and then, when your own scientists, never mind people from the outside, start raising serious scientific concerns on the basis of studies that have been done, the nonscientists, the managers, decide to water down those conclusions. I just don't understand why that would be, unless you guys are under pressure from sources that you shouldn't be, because you are in business not to kow-tow to the manufacturers, or to the surgeons, or to anybody else but, in fact, to protect the well-being of the American public (Weiss, 1990: 153).

ASSESSING THE FDA'S ACTIONS

There are four theoretical frameworks from which to assess the actions of the FDA in the context of the silicone breast implant controversy. Public interest theory maintains that the government should serve as the guardian of the public good. Capture theory maintains that regulations actually serves the interests of those who are to be regulated. Bureaucratic theory views power as beginning inside of regulatory bodies. Finally, economic theory focuses on an assessment of the effectiveness of regulations (Gerston, Fraleigh, and Schwab, 1988).

DISCUSSION QUESTIONS

1. Both rulemaking and litigation occur in the context of uncertainty, in that rules must be adopted and lawsuits resolved even as our state of

scientific knowledge about a particular product or device is evolving. Consider the case of silicone breast implants. At what point in time, if any, should the FDA ideally have intervened with the promulgation of regulations? How might the timing of such regulations have impacted the commencement or continuation of the litigation?

2. Compare and contrast the use of epidemiology in litigation relating to injuries alleged to have arisen from the use of silicone breast implants with the use of epidemiology in the context of agency decisionmaking related to silicone breast implants as a medical device. Consider in your response the extent to which each process (1) is influenced by the media, consumer advocacy groups, professional organizations, and other political influence; (2) considers and incorporates epidemiological findings; and (3) promotes additional epidemiological investigation.
3. Assess the FDA'a actions using each of the four theoretical frameworks discussed above. How do the results of these analyses differ depending on the framework? What are the implications of any existing differences?

REFERENCES

Amid heavy lobbying, government is considering ban on breast implants. (1991). *New York Times*, Oct. 21, A1, A12.

Angell, M. (1992). Breast implants—protection or paternalism. *New England Journal of Medicine, 326*, 1695–1696.

Bolton, D. C. (1992). The evolution of breast implant litigation. In A. M. Levine (Chair). *Litigating Breast Implant Cases—A Satellite Program. Litigation and Administrative Practice Course Handbook Series, Drug and Medical Device Litigation*, PLI Order No. H451, Nov. 12 (pp. 127–236). New York: Practicing Law Institute.

Congressional Quarterly's Federal Regulatory Directory. (1990). Washington, D.C.: Congressional Quarterly Press.

Food and Drug Administration. (1992). Update on Silicone Gel Filled Breast Implants, Press Release, May 27.

Freire, P. (1978). *Pedagogy of the Oppressed.* New York: Penguin.

General and Plastic Surgery Devices. (1982). 47 *Federal Register* 2810, 2820, codified at 21 C.F.R. part 878.

Gerston, L., Fraleigh, C., & Schwab, R. (1988). *The Deregulated Society.* Pacific Grove, California: Brooks/Cole.

Goldberg, M. Z. (1991). FDA panel recommends that breast implants stay on market. *Trial, 27*, 74–77.

Hilts, P. J. (1992). FDA acts to halt breast implants made of silicone. *New York Times*, Jan. 7., A1.

Human Resources and Intergovernmental Relations Subcommittee of the Committee on Government Operations of the House of Representatives. (1993). *The FDA's Regulations of Silicone Breast Implants*, 102nd Congress, 2nd Session, December 1992. Washington, D.C.: Government Printing Office.

Kessler D. (1992). The basis of the FDA's decision on breast implants. *New England Journal of Medicine, 326*, 1713–1715.

Levine, A. M. (1992). Introductory remarks. In A. M. Levine (Chair). *Litigating Breast Implant Cases-A Satellite Program. Litigation and Administrative Practice Course Handbook*

Series, Drug and Medical Device Litigation, PLI Order No. H451, Nov. 12 (pp. 11–13). New York: Practicing Law Institute.

Lloyd, M. (1992). Testimony. General and Plastic Surgery Devices Panel Meeting, 1, Feb. 18–20 (pp. 84–90). Bethesda, Maryland: Food and Drug Administration.

Medical Device Amendments of 1976, 21 United States Code § 360C (effective May 28, 1976), 21 United States Code §§ 301–393 (1976).

O'Keefe, D. F., Jr., Spiegel, R. A. (1976). *An Analytic Legislative History of the Medical Devices Amendments of 1976.* Washington, D.C.: Food and Drug Law Institution.

Weiss, Rep. (1990). Statement to FDA official. In *Is the FDA Protecting Patients from the Dangers of Silicone Breast Implants?* Hearings Before the Human Resources and Intergovernmental Relations Subcommittee of the Committee on Government Operations of the House of Representatives. 101st Congress, 2nd Session.

21 Code of Federal Regulations § 878.3540 (1990).

21 Code of Federal Regulations §§ 897.14, -.30, -.32, -.34 (1996).

CHAPTER 6

Case Study Four

The Regulation of Tobacco

> This vice brings in one hundred million francs in taxes every year. I will certainly forbid it at once-as soon as you can name a virtue that brings in as much revenue (Napoleon III, quoted in Goodman, 1993: 191)
>
> Wayne McLaren, who portrayed the rugged 'Marlboro Man' in cigarette ads but became an anti-smoking crusader after developing lung cancer, has died, aged 51 ... His mother said: 'Some of his last words were: "Take care of the children. Tobacco will kill you, and I am living proof of it".' ... Mr. McLaren, a rodeo rider, actor and Hollywood stuntman ... was a pack-and-a-half-a-day smoker for about 25 years. In an interview last week, Mr. McLaren said his habit 'had caught up with me. I've spent the last month of my life in an incubator and I'm telling you, it's just not worth it" (Guardian, July 25, 1992, quoted in Goodman, 1993: 239).

THE FDA AND THE REGULATION OF DRUGS

An examination of the role of the FDA in the controversy relating to the regulation of tobacco is facilitated by an understanding of the FDA's role in general, as well as the evolution of that role.

The Bureau of Chemistry of the Federal Agriculture Department was created in 1862 and a food and drug law recommended in 1880 (Food and Drug Administration, 1995b). Congress first enacted legislation to regulate drugs with its passage of the Pure Food and Drug Act in 1906. The legislation was limited in scope and focused on preventing mislabeled drugs from reaching the market. This purpose reflected what had been the impetus for the law's passage: the inflated health claims of Dr. Harvey Wiley, who was serving as the head of the Bureau of Chemistry of the United States Department of Agriculture, as well as the unsanitary conditions and practices that were described in Upton Sinclair's *The Jungle* (see Young, 1972). In 1912, Congress passed the Sherley Amendment to define more clearly the FDA's authority.

The subsequent Food, Drug, and Cosmetic Act of 1938 was similarly in response to a tragedy. Sulfanilamide, an antibiotic, was marketed as an elixir by the Massengill Company. The substance proved to be quite toxic and, as a result of its use, over 100 people died. The FDA was unable to intervene except with respect to the labeling of the product: the term "elixir" required that the substance contain alcohol and sulfanilamide did not (Jackson, 1970).

The Food, Drug, and Cosmetic Act permitted the FDA to formulate and adopt a regulation that established the system of prescription-writing, so that the decision making with regard to the use of a substance was transferred from the consumer-patient to the physician. The information upon which the physician relied in making this decision was, however, provided by the pharmaceutical companies. The legislation provided, in addition, for the protection of the consumer (Temin, 1980; Temin, 1979).

Pursuant to the provisions of the 1938 Act, manufacturers were required to file a new drug application (NDA) with the FDA before it could begin marketing the drug. The manufacturer was required to support the NDA with details about the safety of the drug and the procedures used to determine its safety. The FDA would review this information, but did not conduct its own tests for product safety. The NDA would become effective within 60 days, absent an objection from the FDA. At this point, the manufacturer could begin to market the pharmaceutical product.

A Congressional investigation of the pharmaceutical industry was initiated in 1958 in response to concerns about the high cost of drugs and the outrageously high profits within the industry. This investigation was led by Senator Estes Kefauver who, at the time, was the Chair of the Senate Subcommittee on Antitrust and Monopoly. The Kefauver-Harris Amendments of 1962 were passed on the heels of the thalidomide debacle, in which many women who had ingested thalidomide during their pregnancies gave birth to children with limbs that resembles seal flippers (Mintz, 1962). These amendments required that the FDA approve both the safety and the efficacy of new drugs, including all NDAs that had become effective since 1938.

In order to effectuate this efficacy review, the FDA contracted out the function to the National Academy of Sciences (NAS). The NAS formed numerous panels of experts, who were charged with the responsibility of examining data relating to the efficacy of drugs that had been marketed between 1938 and 1962. The panels classified the drugs they reviewed into one of six categories: probably effective, possibly effective, ineffective, ineffective as a fixed combination, and effective, but with qualifications. The FDA reviewed these reports and issues notices, known as DESI (drug effectiveness study implementation) notices to the manufacturers whose products had been classified as less than effective.

A question that is central to this discussion is how "drug" is to be defined, because if a substance is not classifiable and classified as a drug, the FDA may not regulate it as a drug. According to the federal Food, Drug, and Cosmetic Act, the term "drug" refers to

> (A) articles recognized in the official United States Pharmacopoeia, official Homeopathic Pharmacopoeia of the United States, or official National Formulary,

> or any supplement to any of them; and (B) articles intended for use in the diagnosis, cure, mitigation, treatment, or prevention of disease or man or other animals; and (C) articles (other than food) intended to affect the structure or any function of the body of man or other animals and (D) articles intended for use as a component of any article specified in clause (A), (B), or (C) of this paragraph (21 U.S.C. § 321(g)(1)).

TOBACCO AND ITS EFFECTS

A significant number of individuals in the United States engage in smoking behavior. In 1993, approximately 25% of all Americans smoked. This represents a decline in the prevalence of smoking from 1970 (World Health Organization, 1998). Almost 80% of all smokers began to smoke before the age of 16. Numerous explanations have been offered for the apparent decline in the overall prevalence of smoking, including increased educational efforts, the implementation of restrictions on advertising, increases in the sales tax placed on smoking, an increasing number of public health warnings, the implementation and enforcement of restrictions on the sale of tobacco to minors, and the regulation of smoking in public places (Goldman & Katz, 1998; Hersch, 1998; Viscusi, 1998; World Health Organization, 1998; Lessig, 1995; Comprehensive Smokeless Tobacco Health Education Act of 1986, 1994; Viscusi, 1992; Federal Cigarette Labeling and Advertising Act, 1965).

Among adolescents, however, the prevalence of smoking has increased dramatically. In 1995, more than one-fifth of all children in high school smoked, which was a substantial increase in the proportion of high school students smoking in 1992 (World Health Organization, 1998). As many as 70 percent of teen-age smokers become regular smokers before they have reached the age of 18 (United States Department of Health and Human Services, 1994). This upsurge in the prevalence of smoking among adolescents has been attributed to tobacco manufacturers' aggressive marketing campaigns that have specifically targeted youth. The following statements of Claude E. Teague, Jr., formerly the assistant director of research and development at R. J. Reynolds Tobacco Company, in a 1973 memorandum entitled "Research Planning Memorandum on Some Thoughts About New Brands of Cigarettes for the Youth Market," are illustrative of this approach:

> Realistically, if your company is to survive and prosper over the long term, we must get our share of the youth market. In my opinion this will require new brands tailored to the youth market.
>
> Cigarettes should be markets to underage smokers as a cure for the stress ... awkwardness, boredom of being a teenager and as a way of achieving membership in a group, one of the group's primary values being individuality.
>
> For the pre-smoker and "learner" the physical effects of smoking are largely unknown, unneeded or actually quite unpleasant or awkward. The expected or derived psychological effects of smoking are largely responsible for influencing the pre-smoker to try smoking, and provide sufficient motivation during the "learning" period to keep the "learner" going, despite the physical unpleasantness and awkwardness of the period. In contrast, once the "learning" period is over, the

physical effects become of overriding importance and desirability to the confirmed smoker, and the psychological effects, except the tension-relieving effect, largely wane in importance or disappear" (Schwartz, 1995: A2).

Smoking has been identified as the single greatest preventable cause of premature death in the United States. It has been estimated that approximately 20 percent of all deaths in the United States in 1990 were attributable to smoking (Centers for Disease Control, 1993). Smoking was responsible for almost one-third, or 170,000, of all cancer-related deaths in 1995 (American Cancer Society, 1995). Smoking has been identified as the cause of almost 20 % of all deaths in the U.S. due to cardiovascular disease (American Cancer Society, 1995). Smokers are at increased risk, compared to nonsmokers, of aortic aneurysm, atherosclerosis, coronary heart disease, stroke, and heart attack (Newcomb and Carbone, 1992). Smoking has been found to be associated with other adverse health effects, including vascular disease, oral disease, and reproductive consequences (Wald and Hackshaw, 1996).

Even those who do not smoke themselves, but who are exposed to smoke, may suffer adverse health consequences. Passive, or secondhand, smoke has been linked to impaired respiratory health in children and an increased risk of cardiovascular problems (Environmental Protection Agency, 1994; Stanton and Parmley, 1991).

Until quite recently, cigarette manufacturers have maintained that there is inadequate scientific data to establish a causal link between smoking and cancer, or between smoking and a host of other diseases (Sergis, 2001; Glantz & Balbach, 2000; Bero, Slade, Hanauer, & Barnes, 1996). The human studies that have been conducted to date have been observational studies, which attempt to assess the risk of cancer, for instance, among smokers as it compares to nonsmokers. The tobacco companies have maintained, in essence, that the findings from these studies, which have rather consistently indicated an elevated risk among smokers for certain types of cancers and various other diseases, are inadequate to clearly establish a causal association. Apart from such observational studies, experimental research would be required to definitively establish this link. (See chapter 1 for a review of experimental research.) Such research, though, cannot be conducted because it would violate the basic ethical principles governing research: respect for persons, beneficence, nonmaleficence, and justice (see Loue, 2000).

FDA ACTION AND TOBACCO

Tobacco was originally listed as a drug in the 1890 edition of the United States Pharmacopoeia, but was subsequently removed from the 1905 edition and has not been included in the official listing of drugs since that time. Consequently, tobacco was not among the drugs that the FDA was to regulate at the time that the Food and Drug Act of 1906 was passed (Neuberger, 1963). In general, the FDA did not attempt to regulate tobacco prior to the

1980s. In fact, the FDA specifically stated that it would attempt to regulate cigarettes as drugs only if they were promoted as having beneficial effects on the body (Public Health Cigarette Amendments of 1971, 1972). This had occurred previously in the 1950s, in response to the claims of several cigarette manufacturers that cigarettes could be used to reduce body weight and reduce the frequency of respiratory disease (United *States v. 354 Bulk Cartons*, 1959; *United States v. 46 Cartons*, 1953). The Federal Trade Commission (FTC), which had been regulating cigarette advertising since the 1930s, issued guidelines in 1955 that prohibited false claims in cigarette advertising *(In re Philip Morris and Co.,* 1955).

The British researcher Sir Richard Doll published a study in 1952 that reported a link between smoking and cancer (Doll & Hill, 1952). The tobacco industry responded to the public concern generated by this report by introducing filtered cigarettes and by lowering the tar content (Federal Register, 1995). However, even though the tar levels dropped, the nicotine levels have consistently increased.

In 1964, following the publication of the Surgeon General's Report on Smoking and Health, the FTC began proceedings to require that cigarette manufacturers place warning labels on all cigarette packages and in all cigarette advertisements, to advise potential consumers that smoking could affect their health. The FTC claimed that the failure to warn consumers of the smoking-related dangers constituted "an unfair and deceptive trade practice" within the meaning of the Federal Trade Commission Act. Packages were to state: "Cigarette smoking is dangerous to health and may cause death from cancer and other diseases" (29 Federal Register 8325, 1964). Later, in 1969, the Federal Communications Commission (FCC) issued a notice of proposed rulemaking that would have prohibited cigarette advertising on radio and television (34 Fed. Reg. 1959, 1969). Both of these attempts to regulate smoking advertisements were ultimately impeded and obstructed by Congress (see Kelder & Daynard, 1997). Subsequent efforts to further regulate the advertisement and sale of tobacco under the Consumer Product Safety Act similarly failed (Kelder & Daynard, 1997).

The FDA rejected an attempt by the anti-smoking group Action on Smoking and Health in the late 1970s to have the FDA regulate tobacco as a drug. The FDA concluded at that time that there was insufficient evidence to indicate that cigarettes were intended to affect the structure or the function of the body (*Action on Smoking and Health v. Harris*, 1980).

The American Heart Association later petitioned the FDA in 1988 to regulate low tar cigarettes as drugs; the FDA concurred with this position in 1996 (Food and Drug Administration, 1996). In making this determination, the FDA relied on the findings of numerous epidemiological studies that suggested that smoking was responsible for hundreds of thousands of premature deaths each year, as well as a multitude of adverse physical effects. By the time that the FDA rendered this decision, the American Psychiatric Association, the American Psychological Association, and the United States Surgeon General, among others, had concluded that nicotine is an addictive substance. The FDA further noted that tobacco companies had seemed to be aware of the adverse effects of tobacco for some time and, in fact, had intended that nicotine have the stimulating and sedating effects that were associated with tobacco use (FDA, 1996). For instance, as early as 1963, the General Counsel of Brown and Williamson Tobacco Company had written in an internal document that remained undiscovered publicly until a much later date: "We are, then, in the business of selling nicotine, an addictive drug ... " (Federal Register, 1995: 41,611).

In addition to finding that tobacco is a drug, the FDA concluded that tobacco was also subject to FDA regulation as a drug delivery device because "the primary purpose of parts of the cigarette ... is to effectuate the delivery of a carefully controlled amount of the nicotine to a site in the human body where it can be absorbed" (Food and Drug Administration, 1995a). The controlling statute defines a device as

> an instrument, apparatus, implement, machine, contrivance, implant, in vitro reagent, or other similar or related article, including any component, part, or accessory, which is- ...
>
> (3) intended to affect the structure or any function of the body of man or other animals, and which does not achieve its primary intended purposes through chemical action within or on the body of man or other animals and which is not dependent upon being metabolized for the achievement of its primary intended purposes (21 U.S.C. § 321(h), 1994).

This position was not unreasonable, in view of tobacco manufacturers' efforts to manipulate the nicotine levels in the cigarettes. The FDA found, for instance, in its examination of patents filed by various tobacco companies, that the level of nicotine was increased by adding nicotine to the cigarettes' filters and wrappers (Kessler, 1995). Manufacturers increased the nicotine levels in cigarettes with low tar because consumers were known to smoke fewer of these cigarettes. Certain tobacco plants were bred specifically to increase the nicotine levels. The R. J. Reynolds executive Claude E. Teague, Jr. had even written:

> In a sense, the tobacco industry may be thought of as being a specialized, highly ritualized, and stylized segment of the pharmaceutical industry. Tobacco products contain and deliver nicotine, a potent drug with a variety of physiological effects (Federal Register, 1995: 41,617–41,618).

Despite these findings, the FDA decided not to ban the use of tobacco, believing that such a ban would be circumvented through smuggling and black market sales. Additionally, adverse health consequences could befall the individuals who were already addicted to the use of nicotine. Accordingly, it adopted a strategy to prevent the initiation of smoking behavior in children and adolescents by restricting the sale of tobacco products to individuals under the age of 18, requiring that vendors verify purchasers' age with identification, prohibiting the distribution of free samples and the sale of cigarettes through vending machines, and implementing restrictions on the advertising and promotion of cigarettes (21 C.F.R. §§ 897.14, 897.30, 897.32, 897.34, 1996).

Whether smoking regulations should be implemented at all has stimulated significant ethical debate. There is general consensus with the premise that the government has the power to regulate smoking in order to protect the health of the public; what remains unsettled is the scope and implementation of that power (Jacobson and Wasserman, 1997) and how to balance that power with individuals' right to "make life-style choices" (Leichter, 1991: 3). The regulation of the use of tobacco has been justified based upon the harm it causes to others, such as from secondhand smoke; the harm that it may do to smokers who are uninformed of the health-associated risks; the need to protect children; and the need to protect those who are coerced, that is, those who are addicted to nicotine (Pope, 2000). Such regulations are analogous to laws that mandate the use of helmets for motorcycle riders and seat belts for drivers and passengers of automobiles. Nevertheless, the tobacco industry has mocked the need for regulations, maintaining that the regulation of smoking behavior is just the beginning of a slide down the slippery slope, eventually leading to the regulation of alcohol use, butter, and bungee jumping, using exactly the same rationale (Ezra, 1993). An advertisement in Philip Morris Magazine quipped:

> Smoking is a civil right,
> Those who don't should join the fight.
> For if one right does disappear,
> The loss of others may be near ... .
> Too many calories can cause you to die,
> So let's have a ban on apple pie.
> Once a government restricts a right,
> The end will never be near in sight.
> There's a lesson here ... this is no joke,
> I once had a right to smoke! (Mroz, 1987: 29).

DISCUSSION QUESTIONS

1. Consider the timing and context of FDA efforts to regulate tobacco and silicone breast implants. How are those cases of FDA action or inaction similar? How are they different?
2. Both regulatory action and litigation have played a role in the marketing and use of tobacco and silicone breast implants. What forum has most effectively utilized epidemiological data for resolving the issues presented? Which strategy has been most effective in protecting the public's health?

REFERENCES

Action on Smoking and Health v. Harris. (1980). 655 F.2d 236 (D.C. Cir.).
American Cancer Society. (1995). *Cancer Facts and Figures—1995.*
Centers for Disease Control. (1993). Cigarette—smoking attributable mortality and years of potential life lost—1990. *Morbidity and Mortality Weekly Report, 42,* 645–649.

Comprehensive Smokeless Tobacco Health Education Act of 1986, Pub. L. No. 99-252, 100 Stat. 30, codified at 15 U.S.C. §§ 401–4408 (1994).
Doll, R., & Hill, A. B. (1952). A study of the aetiology of carcinoma of the lung. *British Medical Journal, 2*, 1271–1285.
Environmental Protection Agency. (1992). *Respiratory Health Effects of Passive Smoking: Lung Cancer and Other Disorders. Washington, D.C.: United States Government Printing Office.*
Ezra, D. B. (1993). Get off your butts: The employer's right to regulate employee smoking. *Tennessee Law Review, 60*, 905–955.
Federal Cigarette Labeling and Advertising Act, Pub. L. No. 89-92, 79 Stat. 282 (1965), codified as amended at 15 U.S.C. §§ 1331–1340 (1994).
Food and Drug Act of June 30, 1906, 34 Stat. 768 (1906).
Food and Drug Administration (FDA)(1995a). Analysis Regarding the Food and Drug Administration's Jurisdiction Over Nicotine-Containing Cigarettes and Smokeless Tobacco Products. 60 Fed. Reg. 41,453–41,787.
Food and Drug Administration (FDA). (1995b). Fiscal Year 1995 Almanac.
Food and Drug Administration. (1996). Regulations Restricting the Sale and Distribution of Cigarettes and Smokeless Tobacco to Protect Children and Adolescents, 61 Fed. Reg. 44,396, codified at 21 C.F.R. parts 801, 803, 804, 807, 820, 897 (1997).
Glantz, S. A., Balbach, E. D. (2000). *Tobacco War: Inside the California Battles*. Berkeley, California: University of California Press.
Glantz, S. A., Slade, J., Bero, L. A., Hanauer, P., & Barnes, D. E. (1996). *The Cigarette Papers*. Berkeley, California: University of California Press.
Goldman, L., Glantz, S. (1998). Evaluation of antismoking advertising campaigns, *Journal of the American Medical Association, 279*, 772–777.
Goodman, J. (1993). *Tobacco in History: The Cultures of Dependence*. London: Routledge.
Harris, R. (1964). *The Real Voice*. New York: Macmillan.
Hersch, J. (1998). Teen smoking behavior and the regulatory environment. *Duke Law Journal, 47*, 1143–1170.
In re Philip Morris and Co. (1955). 51 F.T.C. 857.
Jackson, C. O. (1970). *Food and Drug Legislation in the New Deal*. Princeton, New Jersey: Princeton University Press.
Jacobson, P. D., Wasserman, J. (1997). *Tobacco Control Laws: Implementation and Enforcement*. Santa Monica, California: Rand.
Kefauver-Harris Amendments of 1962, Pub. L. No. 87-781, 76 Stat. 780 (1962).
Kelder, G. E., Jr., Daynrad, R. A. (1997). The role of litigation in the effective control of the sale and use of tobacco. *Stanford Law and Policy Review, 8*, 63–87.
Kessler, D. (1995). Statement. *Regulation of Tobacco Products (Part 3). Hearings Before the Subcommittee on Health and the Environment of the Committee on Energy and Commerce.* 103d Cong. 5.
Leichter, H. M. (1991). *Free to Be Foolish: Politics and Health Promotion in the United States and Great Britain*. Princeton, New Jersey: Princeton University Press.
Lessig, L. (1995). The regulation of social meaning. *University of Chicago Law Review, 62*, 943–1045.
Loue, S. (2000). *Textbook of Research Ethics: Theory and Practice*. New York: Kluwer Academic/Plenum Publishers.
Mintz, M. (1962). Heroine of FDA keeps bad drug off market. *Washington Post*, July 15, 1.
Mroz, L. C. (1987). Smoking ban? What next? *Philip Morris Magazine*, Summer, 29.
Newcomb, P. A., Carbone, P. P. (1992).The health consequences of smoking: Cancer. *Medical Clinics of North America, 76*, 305–331.
Public Health Cigarette Amendments of 1971. (1972). Hearing on S. 1454 Before the Consumer Subcommittee of the Senate Committee on Commerce, 92d Cong. 240.
Shirley Amendments, Act of August 23, 1912, 37 Stat. 416 (1912).
Schwartz, J. (1995). 1973 cigarette company memo proposed new brands for teens; RJR official cited need for "share of youth market." *Washington Post*, October 4, A2.

Sergis, D. K. (2001). *Cipollone v. Liggett Group. Suing Tobacco Companies.* Berkeley Heights, New Jersey: Enslow Publishers, Inc.

Temin, P. (1980). *Taking Your Medicine: Drug Regulation in the United States.* Cambridge, Massachusetts: Harvard University Press.

United States Department of Health and Human Services. (1994). *Preventing Tobacco Use Among Young People: Surgeon General's Report.* Washington, D.C.: United States Government Printing Office.

Viscusi, W. K. (1998). Constructive cigarette regulation. *Duke Law Journal, 47,* 1095–1131.

Viscusi, W. K. (1992). *Smoking: Making the Risky Decision.* Oxford: Oxford University Press.

Wald, N. J., Hackshaw, A. K. (1996). Cigarette smoking: An epidemiological overview. *British Medical Bulletin, 52,* 3–11.

World Health Organization. (1998). Tobacco or Health: A Global Status Report. Geneva: Author.

Young, J. H. (1972). *The Toadstool Millionaires: A Social History of Patent Medicine in America Before Federal Regulation.* Princeton, New Jersey: Princeton University Press.

29 Federal Register 8,325 (1964).

34 Federal Register 1,959 (1969).

60 Federal Register 41,453–41,683 (1995).

CHAPTER 7

Law, Epidemiology, and Community Organization and Advocacy

> Our greatest glory consists not in never failing, but in rising every time we fail.
>
> RALPH WALDO EMERSON

> Never doubt that a small group of thoughtful and committed citizens can change the world. Indeed it's the only thing that ever has.
>
> MARGARET MEAD

DEFINING ADVOCACY

Advocacy has been defined as

> simply ... speaking up. Each of us have [*sic*] experienced some issue that needs the attention of the larger community. Activism is taking advocacy several steps further, identifying that issue(s), drawing attention to them, and working towards a solution to those issues.... Activism is taking leadership with other people, to empower people in the community, and to make a difference (no matter how small) in their communities (Chang, 1993: no page number).

How, then, do communities become empowered and organized to act?

STAGES OF DEVELOPMENT IN COMMUNITY ORGANIZATION AND ADVOCACY

Braithwaite and colleagues (1994) have depicted the pathway through which a community progresses from disenfranchisement to empowerment and the ultimate improvement of health. The community may be disenfranchised and unempowered as the result of poverty, alienation, racism, sexism, structural unemployment, poor education, self-defeating behaviors, and environmental stress. Through the formulation of community coalitions (identifying leaders, convening meetings, and developing coalition rules), the community is able to embark on numerous foundational activities, with the ultimate goal of advancing its interests. These activities include the conduct of relevant interviews and focus groups, case studies, participant observation, videography, and action research. The results and findings from these and other activities will be used as the basis of resource development, strategic planning, and policy initiatives.

Community groups often organize around specific issues (Labonte, 1994). They may be assisted by health professionals and their institutions in attending to individual-level needs, organizing efforts, coalition advocacy, political action, and group development. Each such sphere addresses a different level of organization: the interpersonal (personal empowerment), the intragroup (small group development), the intergroup (community organization), and the interorganizational (coalition building & advocacy, political action) (Labonte, 1994; Israel, Checkoway, Schulz, and Zimmerman, 1994). Ultimately, then, ideas and direction come from the community, and the professional may serve as a resource and a catalyst in the process (Rifkin, 1995). The initial community mobilization may occur as the result of additional education (Freire, 1970).

Speer and Hughey (1995) have identified four distinct phases in this progression and organization. The first phase consists of assessment, in which critical issues are identified and defined. Often, this occurs in a one-to-one engagement across the community, in an attempt to connect with individuals and to facilitate communication. The second, or research phase, provides a framework for the examination of the roots of the issues that were identified during the assessment phase. During the third, or action, phase, there is a collective attempt to exercise power and to develop strategies for the exercise of that power. Finally, reflection allows participants in the process to examine how the organization has evolved through this cycle and to identify future directions for collective attention and effort.

In building the resulting coalitions, it is critical that each group's constraints, such as its own internal rules, be acknowledged (Chang, 1993). We may better understand how and why disenfranchised communities are able to progress to establish meaningful coalitions by examining the culture of social organizations and movements.

Stage One: Assessing the Needs of the Community

The needs of a particular community can be determined through a needs assessment. A needs assessment is "a systematic set of procedures undertaken for the purpose of setting priorities and making decisions about program or organizational improvement and allocation of resources. The priorities are based on identified needs" (Witkin & Altschuld, 1995: 4). Need can be conceptualized of as "a discrepancy or gap between 'what is,' or the present state of affairs in regard to the group and situation of interest, and 'what should be,' or a desired state of affairs" (Witkin and Altschuld, 1995: 4).

Numerous strategies exist for conducting needs assessments. One of the most common strategies is known as the convergence model. This method relies on multiple levels or tiers of input, which constitute the sources of data, and multiple data collection strategies or tracks, which comprise the instruments. For instance, the 1998 Northeast Ohio HIV/AIDS Needs Assessment, conducted pursuant to the Ryan White Comprehensive AIDS Resources Emergency (CARE) Act of 1991, utilized three tiers and four tracks (Loue, Faust, and O'Shea, 2000). The tiers consisted of (1) secondary data, which included epidemiological data and census data; (2) data from providers, obtained through focus groups and surveys; and (3) data from consumers, obtained through surveys and interviews. The tracks consisted of secondary data (census and epidemiological data), focus groups, surveys, and interviews. These strategies combined to yield the proportion of the target community in various stages of HIV disease, estimates of the proportion of the community infected or affected by HIV, data relating to current use of services, data relating to the need for services that were currently unavailable, sources of care and services, barriers to consumers' receipt of services, systemic changes that were believed necessary, the projected need for services, and funding priorities as perceived by consumers and providers of services. These data allowed, in turn, permitted a better understanding of current fulfilled need, current unmet need, and future need, and provided data critical to a determination of funding priorities (Loue, Faust, & O'Shea, 2000).

Using Epidemiology

Numerous examples exist of the successful use of epidemiology or epidemiological methods in determining a community's needs and the subsequent use of the data to establish goals and objectives for future action. As indicated above, epidemiological data was utilized in the 1998 Northeast OHIO HIV/AIDS Needs Assessment as a basis for estimating the proportion of the community that was infected or affected by HIV/AIDS. The Washington Transgender Needs Assessment Survey provides yet another example of how epidemiology can be utilized. This research was premised on surveys of 263 individuals who self-identified as transgender (Xavier, 2001). (See chapter 10 for a definition of transgender.) One-quarter of the individuals reported that they were HIV-positive, and almost another quarter reported that they did not know their serostatus. More than one-third of the respondents reported suicidal ideation; of those reporting suicidal ideation, almost one-half had made previously attempted suicide. More than one-third reported that their drinking was "problem drinking," and more than one-third reported that they abuse illicit drugs. These findings have helped to make health professionals and policymakers aware of various needs within this transgender community, including the need to develop appropriate

HIV/AIDS prevention education materials, the need for adequate substance abuse treatment programs, and the need to outreach to the transgender community regarding such efforts (Xavier, 2001).

Stage Two: Using Epidemiology for Research

Epidemiologists are rarely mentioned in the context of community organization and advocacy. When a mention of health professionals does appear, the usual focus is on health promotion, which has been viewed "as a response to social movement groups and the challenges they make to government" (Labonte, 1994: 254). However, as indicated above, research and analysis may constitute significant components of advocacy strategy and research is what epidemiologists do best.

Labonte (14) has identified three categories of health problems that may be the focus of advocacy efforts: diseases, such as HIV/AIDS; behaviors, such as unprotected sex and smoking; and social conditions, including poverty, pollution, and discrimination. Labonte has maintained that each set of problems requires a discrete approach to health: the medical, behavioral, and socioenvironmental approaches, respectively. Notwithstanding Labonte's assertion, the ability to address successfully a specified health problem may require action in all three such domains (medical, behavioral, and socioenvironment) and the approaches across such domains may share one or more critical foundational elements.

Epidemiology may constitute one such element. For instance, epidemiology can be used as a tool to identify the incidence and prevalence of a specific disease in the community of interest, to assess the role of a specific behavior in preventing or facilitating the transmission of a specific disease, and to examine the effect of various extraneous factors on the development of a disease or behavior, such as the role of neighborhood violence in the prevalence of violent behavior among adolescents. The role of epidemiology may be best explained by illustration, using environmental justice as an example.

The Environmental Protection Agency has defined environmental justice as

> the fair treatment and meaningful involvement of all people, regardless of race, ethnicity, income, national origin or educational level with respect to the development, implementation, and enforcement of environmental laws, regulations, and policies. Fair treatment means that no population, due to policy or economic disempowerment, is forced to bear a disproportionate burden of the negative human health or environmental impacts of pollution or other environmental consequences resulting from industrial, municipal, and commercial operations or the execution of federal, state, local and tribal programs and policies (Environmental Protection Agency, 1998: 2).

"Health" has been defined as "a state of complete physical, mental, and social well-being and not merely the absence of disease or infirmity" (World

Health Organization, 1986). "Environmental health" refers to "freedom from illness or injury related to exposure to toxic agents and other environmental conditions that are potentially detrimental to human health" (Institute of Medicine, 1995: 15). Environmental justice asserts as its fundamental premise that communities with high proportions of racial or ethnic minorities or of low income members are disproportionately subjected to exposures to numerous environmental hazards and burdens and, as a consequence, are disproportionately experiencing adverse health outcomes (Institute of Medicine, 1999).

Epidemiology may serve as a vehicle by which to assess the health status of the communities involved and to determine the role played by specific environmental factors in that health status. Epidemiology, for instance, can be used as a method to evaluate the role of a specific exposure to a specific outcome, controlling for factors that may confound the true relationship, such as economic status, differential access to health care, race and ethnicity, and daily work and living habits, such as smoking. Sexton's (1996) conceptual model for generating hypotheses about environmental justice issues provides one such example. He illustrates this model by posing the question, "What are the differential effects of class and race on health risk?" (Sexton, 1996: 75). This question necessarily involves classification of individuals by race/ethnicity and by socioeconomic class, which is comprised of income, education, and occupation. He then asks, "How do class and race affect exposure- and susceptibility-related factors?" Exposure-related factors include proximity to sources, occupation, activity patterns, diet, drinking water supply, and tobacco consumption, while susceptibility-related factors include genetic predisposition, health status, nutritional status, age, and gender. Finally, he asks, "How do exposure- and susceptibility-related factors affect health risk." This requires an analysis of emission sources, environmental concentrations, human exposures, internal doses, and adverse effects. Epidemiological methods may be key in this model, helping to understand the independent and dependent effects of race, class, and the exposure- and susceptibility-related factors.

The collection of data by professional epidemiologists may be enhanced and enlarged with the assistance of community-based individuals who assist with the planning, implementation of the research. These same individuals, and others, may later assist with the evaluation of the research and the dissemination of the research findings (Banner, DeCambra, Enos, Gotay, Hammond, Hedlund, Issell, Matsunaga, & Tsark, 1995; Drevdahl, 1995).

Numerous examples of such community participation exist. For instance, Carol Roos, a school district social worker in Tucson, Arizona, spent four months going house to house to investigate the seemingly high rate of occurrence of disease believed to be attributable to exposure to trichlorothylene (TCE), a toxic solvent that had been used by various companies in the area in their aircraft maintenance operations. The study failed to demonstrate a consistent link between exposure and various disease outcomes. However, it did heighten community awareness of the contamination and prompted the community, industry, and government agencies to address the contamination and to seek potential solutions (Brown, 1993). The case of environmental pollution in Woburn, Massachusetts is best known, through the book, *Civil Action* (Harr, 1995) and the subsequent film of the same name, for the court battles that were fought in an effort to establish the resulting injuries and to obtain due compensation. However, members of the Woburn community played significant roles in making their concerns known and in bringing the high rate of illness to the attention of health investigators and attorneys.

Cooperation between epidemiologists and the community may be enhanced if there is mutual trust and respect between the scientist and the community, the scientist knows and understands the community, the scientist and the community have a clear understanding of the policy and programmatic issues the underlie the research, the decisionmaking process is clear, and there is an effort made to develop community expertise and capacity (Lillie-Blanton and Hoffman, 1995).

Stage Three: Developing Strategies in Community Organization and Advocacy

Community organization and advocacy may rely on numerous strategies to accomplish the stated and unstated goals. These strategies may include reliance on the media, attempts to effectuate legal changes through legislative or regulatory proposals by challenging a particular situation or law in court, and attempts to effectuate political change by campaigning for or against officials with specific views on the item of interest. Additional strategies include public education; government lobbying at the local, state, and/or federal levels; and research and analysis (Chang, 1993). Community organization in support of a specific outcome may also be accomplished through action on multiple levels, including the individual, the organizational, the community, and the political levels (Loue, 1996).

Successful advocacy efforts, regardless of the strategy utilized, rest, at least in part, on the ability to identify what is important to the group. Chang (1993) has set forth five aspects to be considered before moving forward:

1. What is the problem to be addressed?
2. How narrow or broad is the goal to be addressed?
3. What are the short- and long-terms plans or goals? Have all concerned parties been included in the planning process?
4. Who or what is needed as a part of the organization to make it effective?
5. Is the plan realistic? Does the issue reflect an actual need in the community?

The most effective community coalitions are those that are holistic and comprehensive, flexible and responsive. They help to build a sense of community, encourage residents to participate in community life, serve as a vehicle for empowering the community, encourage and value diversity, and encourage the development of innovative solutions to existing problems (Wolff, 2001).

Advocacy as Social Movement

Social movements may be thought of as

> a *bundle of narratives*, which when expressed within an interactional arena by participants strengthens the commitment of members to shared organizational goals and status-based identities, sometimes in the face of external opposition (Fine, 195: 128).

Social movements both exist in space and serve as a space for interaction:

> A social movement, whether formalized into a lobbying organization or existing as an informal support or consciousness-raising group, exists "out-there" in the association of civil society, ... ? social spaces" or the informal, unmanaged and associational sectors of society (Labonte, 1994: 225).

As a space, the social movement serves as a "staging area for behavior" (Fine, 1994; Fine, 1981), or as a place of culture enactment (Fine, 1994).

The group, regardless of its size or goals, develops an idioculture through interaction and members' recognition of shared, repeated, meaningful references that ultimately serve to form a collective identity. Accordingly, that idioculture consists of shared knowledge, beliefs, behaviors, and customs. This idioculture serves as a point of reference for group members, provides the basis for further action, and serves to distinguish "insiders" from those deemed to be "outsiders" (Fine, 1994).

The culture of any specific social movement consists of three primary components: identification, rituals, and resources. Identification may be effectuated and enhanced through ideology:

> A political ideology is intended, via action, to establish the identity of a body of persons who are thereafter to be understood to be related to one another in a particular way. The relationship is only one amongst many that each of the potential members of this group may, at a given time, have with a number of other persons, but it is the only relationship which ought, according to the ideology, to embrace them all .... Without commitment the group cannot hope to transform its circumstances with a view to eliminating or isolating relationships incompatible with the one deemed to be ideologically sound (Manning, 1976: 154–155).

Ritual action, even more than talk, serves to enhance the bond between group members and to coalesce them into a group (Fine, 1989; Turner, 1988). The use of narratives in the formation of such rituals is discussed below.

The culture of a movement is a product of its existing resources and also serves as the basis for the solicitation of additional resources. Resources include both material resources and symbolic resources, as well as a communication network by which to coordinate action and a consensual authority system to maintain social control and routinization (Fine, 1995).

Narrative as Strategy

Narratives help to construct a shared meaning and to promote group identity and cohesion (Hardesty & Geist, 1987; Martin, Hatch, & Sitkin, 1983). Often, these narratives address individuals' own life histories and experiences. Fine (1995: 134) has argued that the

> process of exemplifying a frame occurs through the stories that members share, through the collective bundle of narratives that are treated as relevant to the movement ideology.... A social movement is not only a set of beliefs, actions, and actors, but also a *bundle of stories.* Movement allegiance depends on personal accounts, which concretely clarify that extended effort is worthwhile and that others have similar experiences and feelings.

The use of the narrative rests on the assumption that there exists both a performer (the narrator of the story) and an audience (the listener(s)). Three types of narratives have been identified: (1) horror stories, (2) war stories, and (3) stories with happy endings.

Horror stories relate affronts to the narrator and serve to promote active involvement with the movement. These negative experiences permit the narrating individuals to address their public stigma, thereby transforming the public deficit to a subcultural advantage (Fine, 1995), and persuading the audience that is presumed to be sympathetic (Rice, 1992). These narratives, which may portray a "social problem" that demands remedial action, may evoke both anger and sympathy among audience members and, consequently, serve as a call to action (Fine, 1995). Examples of such horror stories include the narratives of therapy patients who allege that their therapists persuaded them during the course of therapy that they had been molested, when they had, in fact, not been.

In contrast, war stories relate situations that members have faced in the context of their activities in and for the movement. Such narratives underscore the value of community through the sharing and ratification and incorporation into the group culture of events deemed to be meaningful. Examples of war stories include those relating to public responses to demonstrations and counter-demonstrations.

Narratives with happy endings relate unexpected benefits or rewards, thereby persuading members that support for their cause does exist. Such stories may help to boost participant morale.

The content of the movement ideology affects the nature of the narratives that are used. Personal narratives, for instance, are more common in rights-oriented movements as compared to policy-oriented movements. Personal narratives serve to establish a connection between the storyteller and the audience, through either of two strategies. They may attempt to achieve either acceptance of the blemished, and subsequently saved or healed, self. This is best exemplified by the accounts of participants in Alcoholics Anonymous, who often relate incidents of personal failure and the role of the organization or movement in their redemption.

The second strategy, that of stigma-deflection, attempts to deflect the stigma that would otherwise adhere to the narrator, portrayed as an unblemished figure. This strategy is best illustrated by the accounts of members of Victims of Child Abuse Laws (VOCAL), an organization founded in Minnesota in 1984 in response to numerous legal actions against parents for the alleged abuse of their children. Such accounts are often horror stories, portraying the storyteller as a victim of an unjust system that warrants and demands systemic changes, including changes in the relevant laws (Fine, 1995).

The Use of the Media

Media advocacy has been defined as

> an effort to move public discourse from a focus on individual blame to a more proper focus on societal conditions and institutional arrangements that are at the root of public health problems. Ultimately, media advocacy is a source of power for citizen groups to make their concerns known and to build support for changes in public policy (Russell, Voas, DeJong, and Chaloupka, 1995: 240).

Public discourse involves more than merely the events that are reported by the news, and may include such things as slogans and movies (Gamson, 1995). The corresponding images may have no fixed meanings. Rather, their meanings result from a process of negotiation with the members of a diverse audience that bring with them various personal associations and experiences. This negotiation process may result in audience interpretation and assignment of a meaning that differs from that which was originally intended (Gamson, 1995).

Media and Framing

The media may play a critical role in inspiring and/or legitimating a particular issue or social movement through its framing of the issue or movement. Three components—injustice, agency, and identity—comprise collective action frames, defined as "action oriented sets of beliefs and meanings that inspire and legitimate social movement activities and campaigns" (Snow & Benford, 1992; Gamson, 1992).

Injustice. An injustice frame involves the portrayal of actors as having brought about, at least in part, the ensuing harm and suffering. The named target might include, for instance, a corporation or "the system." Too specific a focus may negate the possibility of needed, broader action, while too general a frame focus, such as the amorphous "system," may dilute and defuse feelings of anger and indignation that are critical to the effectuation of collective action (Fine, 1995).

Media often utilizes narratives with an injustice frame. As one media personality stated, "Every news story should, without any sacrifice of probity or responsibility, display the attributes of fiction, of drama" (Epstein, 1973: 241) and should be organized around conflict and problem.

Agency. Media serves to validate the importance of the actors through its provision of attention. The audience is, in essence, told that the group or the actors involved are to be taken seriously. The arrest or suppression of an actor or actor serves to underscore the severity of the situation that is posed.

Many ordinary citizens may not feel a sense of agency, an ability to effectuate change, because they must devote the majority of their time and energy to maintaining their daily lives (Gamson, 1995). Additionally, few individuals are actually presented with an opportunity to participate in an activity through which they can effectuate change (Flack, 1988). The media may further limit public participation because its reliance on the pronouncements of established leaders may reinforce the feeling of individuals that they cannot make a difference (Bennett, 1988).

The media has, however, the ability to impact citizen inaction and citizens' disbelief in their own ability to alter the conditions of their daily lives through their own actions. The media can help to effectuate this transformation in attitude and (in)action through the depiction of successful citizen action on various issues and the portrayal of the citizen-actors as the subject of what is happening (the actors are doing), rather than as the passive objects of events (the actors are being acted upon) (see Gamson, 1995).

Identity. Collective action implies self-identification as part of a "we," distinguished from the "they." How the "we" is defined may evolve and change over time (Melucci, 1989). The collective identity, the "we," may encompass three different layers, which may be separable or may be partially or totally overlapping: the organization, the movement, and the solidary group (Gamson, 1995). The focus at each layer is on the sociocultural, rather than the individual, level. This identity is expressed through various symbols and particularized language, such as a specific style of dress or code words (Gamson, 1995). Mentally ill individuals, for example, may wish to promote an understanding of their situation in conjunction with a particular organization and as part of a larger movement to assure the rights of mentally ill persons.

Identity frames may be "aggregate" or may embody "collective action." Those that are aggregate are not linked to an identifiable "we" and tend to address abstract concepts such as worldwide hunger or injustice. In contrast, those that are characterizable as having collective action identities often have clearly delineated targets, the "they" to which they stand in opposition (Gamson, 1995). Such collective action identities represent a critical element in consensus movements, defined as "organized movements for change that find widespread support for their goals and little or no organized opposition from the population or geographic community" (Wolfson, 1990). The media may represent an adversary to a consensus movement or, alternatively, may become an ally by virtue of its ability to enlarge the audience to which the movement's message is directed, and to sympathetically portray the cause that the movement is championing or the actors who are its advocates.

Strategies for Using the Media. Various strategies for working with the media have been identified that appear to increase the success of advocacy efforts. These techniques include (1) timing press releases and events to enhance the likelihood of broad dissemination of information (Russell, Voas, DeJong, & Chaloupka, 1995; Chang, 1993); (2) simplifying, to the extent possible, the information that is to be conveyed to the intended audience (Russell, Voas, DeJong, & Chaloupka, 1995); (3) making the story "newsworthy" by making it timely, with a clear message, a human dimension, and a central conflict (Chang, 1993); (4) utilizing various media forms, such as television, radio, and newspaper articles, as well as editorials and letters to the editor; and (5) creating press releases that are concise and accurately portray the relevant facts (Chang, 1993).

EPIDEMIOLOGY, ADVOCACY, AND ETHICS

Concerns have been voiced regarding the ethics of an epidemiologist's participation as an advocate:

> The epidemiologist can have several roles in the policymaking process which may include producing the data and its interpretations, presenting specific policy options and projecting the impact of each, developing specific policy proposals and evaluating the effect of policies after they have been implemented. An important question which arises, is can an epidemiologist be both a researcher and an advocate for a specific policy? I believe this question urgently needs to be addressed. Is our scientific credibility adversely affected by our taking a strong advocacy position regarding specific public policies? Even if we believe the answer is "yes" we should also ask whether it is ethical not to advocate (Gordis, 1991: 12S).

Still others have asserted vehemently that epidemiologists can and should assume an advocacy role:

> Epidemiologists are as susceptible to human frailties as any other kind of scientist. In common with all other scientists, they must uphold the highest standards of scientific honesty, integrity and impartiality. At times a problem can arise in relation to the last of these three qualities, impartiality: more than many other branches of biomedical science, epidemiology aims explicitly at detecting and correcting risks to health. Dealing effectively with risks to health often requires advocacy—the opposite of impartiality. Can an epidemiologist be simultaneously an impartial scientist and a passionate advocate? This is a difficult balance! What is at issue is not impartiality but sound scientific judgment. It is possible to be an advocate while at the same time preserving sound scientific judgment (Last, 1991: 171–172).
>
> [W]e have an educational and interpretive function that need not compromise our credibility and indeed will only enhance it. An additional consideration is that since our data have important societal implications, if we want society to

> continue to support our efforts we will have to demonstrate the value of our research for the health of the public. This can only be done if we broaden our responsibility from the research only role to that of policy-related functions (Gordis, 1991: 12S).

Indeed, the responsibility of the epidemiologist to the larger community, separate and apart from the research participants, has been noted in international ethical guidelines. These guidelines implicitly recognize, as well, the appropriateness of an advocacy role:

> Part of the benefit that communities, groups and individuals may reasonably expect from participating in studies is that they will be told of findings that pertain to their health. Where findings could be applied in public health measures to improve community health, they should be communicated to the health authorities.... Research protocols should include provision for communicating such information to communities and individuals. Research findings and advice to communities should be publicized by whatever suitable means are available... (CIOMS, 1991: Guideline 13, at 14).

DISCUSSION QUESTIONS

1. What role can epidemiology play in the formulation of narratives to be utilized in an advocacy effort? In what advocacy efforts, in addition to environmental justice, has epidemiology played such a role?
2. How has epidemiology been used in the media in advocacy efforts and social movements? Identify two such instances and explain why the use of epidemiology in these situations was successful or unsuccessful. How should "success" in this context be defined?

REFERENCES

Banner, R. O., DeCambra, H., Enos, R., Gotay, C., Hammond, O. W., Hedlund, N., Issell, B. F., Matsunaga, D. S., & Tsark, J. A. (1995). A breast and cervical cancer project in a Native-Hawaiian community: Wai'anae cancer research project. *Preventive Medicine, 24*, 447–453.

Bennett, W. L. (1988). *News: The Politics of Illusion*. New York: Longman.

Braithwaite, R. L., Bianchi, C., & Taylor, S. E. (1994). Ethnographic approach to community organization and health empowerment. *Health Education Quarterly, 21*, 407–416.

Brown, P. (1993). Popular epidemiology challenges the system. *Environment, 38*, 16–41.

Chang, R. (1993, July 26). *In Many Tongues: Advocacy Training 101*. San Francisco, California: Asian/Pacific AIDS Coalition (A/PAC).

Council for International Organizations of Medical Sciences (CIOMS). (1991). *International Guidelines for Ethical Review of Epidemiological Studies*. Geneva: CIOMS.

Drevdahl, D. (1995). Coming to voice: The power of emancipatory community interventions. *Advances in Nursing Science, 18*, 13–24.

Environmental Protection Agency, Office of Federal Activities. (1998). *Final Guidance for Incorporating Environmental Justice Concerns in EPA's NEPA Compliance Analyses*. Washington, D. C.: United States Government Printing Office.

Epstein, E. J. (1973). *News from Nowhere*. New York: Random House.

Fine, G. A. (1981). Friends, impression management, and preadolescent behavior. In S. R. Asher, & J. M. Gottman (Eds.). *The Development of Children's Friendships*. New York: Cambridge University Press.

Fine, G. A. (1989). The process of tradition: Cultural models of change and content. In C. Calhoun (Ed.)., *Studies in Comparative Historical Sociology*. Greenwich, Connecticut: JAI Press.

Fine, G. A. (1995). Public narration and group culture: Discerning discourse on social movements. In H. Johnston, & B. Klandermans (Eds.). *Social Movements and Culture: Social Movements, Protest, and Contention* (pp. 127–143). Minneapolis: University of Minnesota Press.

Flack, R. (1988). *Making History*. New York: Columbia University Press.

Flick, L. H., Reese, C. G., Rogers, G., Fletcher, P., & Sonn, J. (1994). Building community for health: Lessons from a seven-year-old neighborhood/university partnership. *Health Education Quarterly, 21*, 369–380.

Freire, P. (1970). *Pedagogy of the Oppressed*. New York: Seabury Press.

Gamson, W. A. (1995). Constructing social protest. In H. Johnston, B. Klandermans (Eds.). *Social Movements and Culture: Social Movements, Protest, and Contention* (pp. 85–106). Minneapolis: University of Minnesota Press.

Gamson, W. A. (1992). The social psychology of collective action. In B. Klandermans, H. Kriesi, S. Tarrow (Eds.). *International Social Movement Research: From Structure to Action*. Greenwich, Connecticut: JAI Press.

Gordis, L. (1991). Ethical and professional issues in the changing practice of epidemiology. *Journal of Clinical Ethics, 44*, 9S–13S.

Hardesty, M., & Geist, P. (1987). Stories of choice and constraint in the pursuit of quality medical care. Paper presented to the Society for the Study of Symbolic Interaction, Urbana, Illinois. Cited in Fine, G.A. (1995). Public narration and group culture: Discerning discourse on social movements. In H. Johnston, B. Klandermans (Eds.). *Social Movements and Culture: Social Movements, Protest, and Contention* (pp. 127–143). Minneapolis: University of Minnesota Press.

Institute of Medicine. (1995). *Nursing, Health, and the Environment: Strengthening the Relationship to Improve the Public's Health*. Washington, D.C.: National Academy Press.

Institute of Medicine. (1999). *Toward Environmental Justice: Research, Education, and Health Policy Needs*. Washington, D.C.: National Academy of Sciences.

Israel, B., Checkoway, B., Schulz, A., Zimmerman, M. (1994). Health education and community empowerment: Conceptualizing and measuring perceptions of individual, organizational, and community control. *Health Education Quarterly, 21*, 149–170.

Labonte, R. (1994). Health promotion and empowerment. Reflections on professional practice. *Health Education Quarterly, 21*, 253–268.

Last, J. (1996). Professional standards of conduct for epidemiologsts. In S. S. Coughlin, T. L. Beauchamp (Eds.). *Ethics and Epidemiology* (pp. 53–75). New York: Oxford University Press.

Last, J. M. (1991). Epidemiology and ethics. *Law Medicine & Health Care, 19*, 166–173.

Lillie-Blanton, M., & Hoffman, S. C. (1995). Conducting an assessment of health needs and resources in a racial/ethnic minority community. *Health Services Research, 30*, 225–236.

Loue, S., Faust, M., & O'Shea, D. (2000). Determining needs and setting priorities for HIV-affected and HIV-infected persons: Northeast Ohio and San Diego. *Journal of Health Care for the Poor and Underserved, 11*, 77–86.

Loue, S., Lloyd, L. S., & Phoombour E. (1996). Organizing Asian Pacific Islanders in an urban community to reduce HIV risk: A case study. *AIDS Education and Prevention, 8*, 381–193.

Manning, D. J. (1976). *Liberalism*. New York: St. Martin's Press.

Martin, J., Feldman, M. S., Hatch, M. S., & Sitkin, S. B. (1983). The uniqueness paradox in organizational stories. *Administrative Science Quarterly, 28*, 438–453.

Melucci, A. (1989). *Nomads of the Present: Social Movements and Individual Needs in Contemporary Society*. Philadelphia: Temple University Press.

Rifkin, S. (1985). *Health Planning and Community Participation*. London: Groom Helm.

Rice, J. S. (1992). Discursive formation, life stories, and the emergence of codependency: 'Power/knowledge' and the search for identity. *Sociological Quarterly, 33*, 337–364.

Russell, A., Voas, R. A., DeJong, W., & Chaloupka, M. (1995). MADD rates the states: A media advocacy event to advance the agenda against alcohol-impaired driving. *Public Health Reports, 110*, 240–245.

Sexton, K. (1996). Environmental justice: Are pollution risks higher for disadvantaged communities? *Health and Environment Digest, 9,* 73–77.
Snow, D. A., & Benford, R. D. (1992). Master frames and cycles of protest. In A. Morris, C. M. Mueller (Eds.). *Frontiers on Social Movement Theory.* New Haven. Connecticut: Yale University Press.
Speer, P. W., & Hughey, J. (1995). Community organizing: An ecological route to empowerment and power. *American Journal of Community Psychology, 23,* 729–748.
Turner, J. (1988). A Theory of Social Interaction. Stanford, California: Stanford University Press.
Witkin, B. R., & Altschuld, J. W. (1995). *Planning and Conducting Needs Assessments: A Practical Guide.* Thousand Oaks, California: Sage.
Wolff, T. (2001). Community coalition building-Contemporary practice and research: Introduction. *American Journal of Community Psychology, 29,* 165–172.
World Health Organization. (1986). Constitution. In World Health Organization. *Basic Documents.* Geneva, Switzerland: World Health Organization.

CHAPTER 8

Case Study Five

Alcohol and Drunk Driving

Wine and beer are alcoholic. This makes them as "cultural," as dependent upon civilized control and organization as bread, in that human beings have to work hard and long... Care, planning, technology, and organization are required for both. Alcohol is to be treated with respect, not because it is "the staff of life" like bread, but because it is exactly the opposite: it gives pleasure, but it is usually unnecessary and potentially dangerous. In ancient Greek myth, wine was a "late-comer" in human history, which meant among other things that people could live without it. Drinking it induced religious awe and direct acquaintance with Dionysius, the god of the vine, of ecstasy; of the group acting as one, of the loss of individual identity (Visser, quoted in Murdock, 1998).

ALCOHOL AND DRUNK DRIVING

The Physiological Effects of Alcohol

Alcohol is absorbed into the bloodstream, primarily from the small intestine. Principal effects of alcohol ingestion include depression of the central nervous system. Many states have imposed a legal driving level of 100 milligrams per deciliter of blood, or less; this standard is often used to define intoxication (Berkow, 1992).

Individuals who drink large amounts of alcohol on a repetitive basis may become tolerant of the effects of alcohol, so that intoxication does not occur with the ingestion of the same amounts. Withdrawal, when it is mild, may include symptoms such as tremor, weakness, and sweating. Long-term effects of chronic, heavy ingestion may include cirrhosis of the liver, peripheral neuropathy, gastritis, and pancreatitis.

The Characterization of Alcohol Use

Alcohol was brought to the English colonies with the very first settlers. Alcohol was used as a medicinal remedy for such ailments as headaches and infections (Murdock, 1998) and as a commercial asset in trading with the Indians (Mancall, 1995). Colonial indulgence of alcohol was heavy by current standards, averaging approximately seven shots a day. Moralists, including Puritan clerics, relied on Biblical passages to warn the colonists of the horrors that would befall them as the result of excessive indulgence: Noah's embarrassment at being found naked and inebriated by his children, and tables covered in vomit after too much wine (Mancall, 1995). Despite such admonitions, however, there were no serious efforts made to control the use or sale of alcohol until after the Revolution.

The much-later movement for Prohibition and temperance was due, at least in part, to the belief that abstinence provided a form of salvation for women from poverty, domestic violence, and abandonment (Murdock, 1998). During the mid- to late 1800s, an increasing number of states adopted dry laws, which prohibited the commercial manufacture, sale, and public consumption of alcohol. Significant concern was expressed regarding women's drinking, as inebriety was believed to be an inherited trait that became worse with each succeeding generation. Sterilization was proposed as a solution to women's drinking.

During the early 1900s, women began to assert their right to drink outside of the home. While women had at first championed prohibition as a mechanism for their protection, some women's groups now began to campaign actively in favor of the repeal of Prohibition. Federal prohibition was ultimately repealed in 1933 (Murdock, 1998).

Popular magazines shifted from their moralistic view of alcoholism in the early part of the 20th century to a more "naturalistic" approach in the 1960s (Linsky, 1971). Accordingly, alcoholism, which was once attributed to internal biological and/or psychological causes, was explained by reference to external causes. McKenzie and Giesbrecht (1981) noted a shift in public emphasis in the 1950s on alcohol's impact on private life to issues related to drunken driving in the 1960s, with a later focus in the 1970s on the economic impact of drinking.

Media coverage of alcohol-related issues has similarly shifted over time. Hingson and colleagues (1988) found that the number of articles related to driving while under the influence of alcohol rose from fewer than 20 in 1980 to greater than 150 in 1983 and 1984. Lemmens and colleagues (1999) found from their review of articles in 5 national newspapers during the period from 1985 to 1991 that articles addressing alcohol have increasingly emphasized public health issues, have de-emphasized clinical aspects of alcoholism, and have shifted towards an emphasis on external environmental factors to define or explain alcohol-related behavior.

Alcohol consumption patterns have, for instance, been found to be associated with cultural norms, rather than with physical symptoms. Greater levels of acculturation among immigrants, for instance, have been found to be associated with current drinking (Li & Rosenblood, 1994), while educational level and occupational status have been found to be correlated with alcohol-related problems (Nawakami, Haratani, Hemmi, & Araki, 1992).

The Epidemiology of Driving-Related Injuries and Deaths

According to statistics of the National Highway Traffic Safety Administration, 17,126 persons were killed in alcohol-related accidents in 1996. This constitutes 40.9% of all people killed in traffic accidents that year. If one counts only those deaths in which one or more persons had a blood alcohol content above the legal limit of 0.1% at the time of the accident, the figure of alcohol-related deaths is reduced to 13,395, or approximately 32% of all fatal crashes (Barr, 1999). In comparison, approximately 15% of total road deaths in Great Britain involve accidents in which one or more people have a blood alcohol content (BAC) over the legal limit of 0.08.

Discussions focusing on the reduction of alcohol-related traffic deaths have generally emphasized a reduction in the allowable BAC to 0.8 and a uniform adoption of a minimum drinking age of 21. However, the adoption of 21 as the minimum age has met with some opposition due, in part, to the existence of "blood borders," which allow individuals under the age of 21 to drive into parts of Canada and Mexico, where the drinking age is 18 (Barr, 1996). Levy and Asch of Rutgers University concluded from their study that:

> It does not appear that the high fatality risk presented by new drinkers can be ameliorated by raising the legal drinking age ... The problem arises not because we permit people to drink when they are "too young," but rather because we permit them to experience the novelty of "new drinking" at a time when they are legally able to drive. If drinking experience preceded legal driving, a potentially important lifesaving gain may follow (Quoted in Barr, 1996: 279–280).

The National Highway and Traffic Safety Administration has estimated that approximately two-thirds of traffic deaths are associated with aggressive driving. Aggressive driving behavior and the use of cell phones may also be related to vehicular injury (Cellar, Nelson, and Yorke, 2000; Irwin, Fitzgerald, and Berg, 2000). Despite this research, however, the majority of individuals believe that most traffic injuries and fatalities are attributable to drunken driving (Barr, 1996).

MOTHERS AGAINST DRUNK DRIVERS

Reinarman (1988: 91) has argued forcefully that the anti-drunk driving movement

> did not spring from any rise in the incidence or prevalence of drinking-driving or in accidents thought to be related to it. In fact, the rate of road accidents in the United States remains lower than in most other Western industrial democracies. It is widely believed that people who drink and drive end up in accidents in which there is a tragic and costly loss of life, limb, and property. However, none of the organizations or leaders of the movement against drinking-driving have even suggested that their efforts were prompted by some sudden rash of drinking-driving accidents. On the contrary, all claim that their work arose from the fact that the injustices attributed to drinking-driving laws have long been a problem and have never been treated seriously by legislatures and courts.

Accordingly, then, if an association between driving while drunk and a loss of life and/or limb does not provide the foundation for success of the anti-drinking-driving movement, what does? The answer may well lie in the *raison d'être* of organizations such as Mothers Against Drink Drivers (MADD) and its claim to moral authority.

The Claim to Moral Authority

MADD came into being as a nonprofit organization in August 1980, largely through the single-minded efforts of Candy Lightner, who had lost her 13-year old daughter as the result of a car accident caused by a drunk driver. At the time of the accident, the driver was on probation for previous DUI (driving under the influence) convictions and had been released on bail, posted by his wife, for another hit-and-run DUI offense that had occurred several days prior to the accident involving Cari Lightner (Reinarman, 1988). The organization was funded with the proceeds from Cari's insurance settlement, Candy Lightner's own savings, and various small grants from the American Council on Alcohol Problems, the National Highway Traffic Safety Administration, and the Levy Foundation (Reinarman, 1988).

From its inception, MADD portrayed itself as the voice of the victim. Weed (1990) has identified three categories of victim status reflective of MADD membership, with each type of victimhood giving rise to a different status: the individually harmed victim, the bereaved victim, and the general community activist.

The Individually Harmed Victim

The individually harmed victim is one who survived an accident caused by a drunken driver. Accordingly, the individual is deemed to possess "experiential knowledge" about drunk driving and can claim a special understanding and expertise on that basis (Weed, 1990). This attributed enhanced understanding and expertise can provide the foundation for leadership, status, and authority in making decisions (Borkman, 1976). In this role, the survivor can express anger, fear, bitterness, and frustration about the event that led to his or her victimhood (Wortman, 1983). One such self-identified victim explained:

> I am not on crutches. I am not in a wheelchair. I do not have brain damage. I see, hear, feel and think as well as ever. In fact, I have no remaining visible injuries. But I do bear the scars inside as the victim of a drunk driving crash.
>
> Because I am President of MADD, most people assume that one of my children was killed in a drunk driving crash. I can never pretend to know the pain of having a child killed. But I do know the pain of being an injured victim (Sadoff, 1989: 4).

The Bereaved Victim

Simos (1979: 227–228) has explained that activism may arise from loss because "self-esteem can be sought by identification with a social movement or cause from which one derives reflected power and glory." Weed (1990: 461) has explained further:

> In an organizlation like MADD this requires that a person play the role of the victim in the name of the deceased. Like the individually harmed victim, the bereaved victim must publically [*sic*] express emotions of loss to demonstrate "experiential knowledge" about the harm drunk driving causes. The bereaved victim, like the harmed victim, can claim to have experiential expertise about the suffering of families which serves as a basis of authority for speaking out on the problem. There is a difference in the focus of bereaved victims' role because they must act in the name of the deceased to "help others" by promoting projects that are seen as preventing future drunk driving fatalities (see also Borkman, 1976).

MADD's own literature describes such a victim:

> You, your son, your daughter, or some other close loved one has been injured or killed by a drunk driver. You're not alone. As a victim, you may now find yourself in the same situation as many others do; feeling overwhelmed and helpless with grief, anger and outrage in reaction to the violent and unexpected death of a loved one.
>
> As the victim of a crime, it is important to know what you can do to tale action to insure the thorough prosecution of the drunken driver so as to lessen the chance that your personal tragedy will be repeated (MADD, 1985: 3).

The moral authority of the bereaved victim is increased in situations in which the deceased individual can be perceived as an "innocent victim." Research has indicated that attaining the status of "innocent victim" is relatively difficult because observers generally do not identify with the suffering experienced by the victim and, instead, will tend to focus on any perceived negative attributes, thereby relieving themselves of any responsibility or guilt (Lerner, 1980). "Innocent victim" status of the deceased, or the injured, is more likely to be granted with the existence of five situational characteristics: (1) the victim is, or is perceived to be, weak; (2) the victim was in the process of doing something respectable; (3) the victim could not possibly be blamed for where he or she was at the time of the supposedly tragic occurrence; (4) the offender is perceived as "bad"; and (5) there was no connection between the victim and the offender (Christie, 1986). As Weed (1990: 462) has explained:

> The extent to which people assign an event these characteristics, motives, and actions in their interpretation determines the extent to which individuals are seen as being innocent victims. As the situation diverges from these characteristics people are free to assign more culpability to the victim.

In the case of Cari Lightner, the victim was a young 13-year old girl, walking on the side of the road to a neighborhood church carnival. She was hit from behind by a drunken driver with a previous drunken driving conviction who, prior to the time of the accident, had had no connection whatsoever to Cari Lightner or to anyone in her family (Weed, 1990; Reinarman, 1988).

The General Community Activist

The general community activist is one who is more disinterested than either the individually harmed victim or the bereaved victim. There may be no individual connection to the specific cause. Rather, the individual's commitment may stem from the belief that a particular activity or organization represents a "good cause," a conviction that community involvement is a key to the resolution of social problems and the restoration of justice (Knoke, 1988), and/or the perception of others as victims requiring assistance (see Lerner, 1980). Historically, middle class women have been more likely than those of other socioeconomic statuses to focus on alcohol use and to be view it as a potential threat to the moral fabric of home and community (see Bordin, 1981). The general activist may portray the drunken driver as an individual whose lack of responsibility ultimately constitutes a danger to all in the community, thereby necessitating the criminalization of such conduct (Rock, 1973; Duster, 1970).

Goals and Strategies

The 1984 chapter organizing materials of MADD focused on three primary areas of activity: public awareness, legal advocacy, and victim assistance:

> 1. *Public Awareness and Education*: Chapters must educate communities about the seriousness of driving under the influence and to the fact that Americans are individually responsible for their decision to drive while intoxicated.... Community awareness and education includes working with the media (newspapers, television, radio, magazines), Speaker's Bureau programs, poster contests, and annual candlelight vigil and educational programs designed for school-age children and youth. 2. *Legal Advocacy*: Tough laws are effective as a deterrent to driving under the influence, only when these laws are enforced.... MADD members are encouraged to educate criminal justice system personnel and lawmakers of the need for specific types of law, consistent enforcement, and swift and certain punishment of offenders. 3. *Victim Assistance*: As the major goal of MADD is to provide assistance to victims and their families, the chapter must be prepared from the beginning to offer one-on-one support to victims who contact the chapter.... The [MADD] volunteer will make appropriate referrals to appropriate community resources and give the family materials such as the Victim Information Packet and grief brochures.... Volunteers will know enough about the legal process to enable victims to be assertive in knowing their rights through the court case (MADD, 1984: 1–2).

Unlike its competitors, such as the Alliance Against Intoxicated Motorists (AAIM), Boost Alcohol Consciousness Concerning the Health of University Students (BACCHUS), Remove Intoxicated Drivers (RID), and Students Against Driving Drunk (SADD), Mothers Against Drunk Driving was willing to accept funding from the alcohol and broadcasting industries and did not advocate increases in alcohol consumption as a means of reducing alcohol consumption or financing treatment services. Rather, MADD's agenda specifically emphasized individual responsibility, the private moral choice of drinkers, and the resolution of issues through the self-regulation of drinkers, the alcohol industry, the media, and advertisers (Reinarman,

1988). Accordingly, MADD's efforts were supported by both the alcohol industry and the media:

> Television began this groundswell by giving airtime to MADD's painful [Congressional] testimony.... It did so not merely because of its perceived importance—important but complex and boring testimony is given all the time without a dream of TV coverage—but because it was emotional, sentimental. No sane news director will pass up a grieving, sobbing mother; it is the basic image of tragedy on which TV thrives.... It was the beginning of an orgy of attention. In their search for safe issues on which to take a 10-second position, TV editorial directors pounced on drunk driving as if it were an end-zone fumble. They called repeatedly, almost weekly for "stiffer penalties".... It was heaven-sent: instantly graspable and without opposition.... At the same time, the National Association of Broadcasters organized a massive public service campaign on the problem... organized a massive public service campaign on the problem.... Broadcasters aren't stupid; the motivation behind all this attention was forestalling any efforts to ban ads [relating to alcoholic beverages] (Freund, 1985: 1).

MADD'S orientation, with its emphasis on individual responsibility and refusal to address systemic aspects of drunken driving, was in harmony with the policies and rhetoric of the agenda of both Reagan and the New Right. Consequently, the timing of the movement contributed to its success (Marshall and Oleson, 1994; Reinarman, 1988). MADD used its political credibility to achieve major successes: the elimination of plea bargaining for drunken driving offenses, the institution of mandatory jail sentences, the reclassification of alcohol related injuries and death accidents to felonies, the development and implementation of "dram shop" (server) liability laws, the institution of random sobriety checkpoints, and the adoption of mandatory treatment laws and of 21 as the minimum drinking age (Reinarman, 1988).

In order to advance its policy agenda, MADD designed a "Rating the States" (RTS) Program to publicize the efforts of each of the states to combat alcohol-impaired driving. As part of this 1993 program, MADD rated each state in the following areas: gubernatorial leadership, statistics and records, enforcement, administrative and criminal sanctions, regulatory control and availability, legislative efforts, prevention and public awareness efforts, youth issues, self-sufficiency programs, innovative programs, and victim issues. Each state was assigned a grade, ranging from F (the lowest) to A (the highest). Publication of the report often led to renewed efforts by the sate to address alcohol-impaired driving (Russell, Voas, Dejong, & Chaloupka, 1995). Several factors have been identified that were critical to the success of this program: (1) the high degree of credibility already associated with MADD as an organization; (2) the high degree of interest that the public generally has in knowing how their own state compares with others; (3) the use of a rating system from A to F that is familiar to almost everyone; (4) a focus on specific political leaders, resulting in political controversy and increased attention and action; and (5) reliance on outside consultants with expertise in marketing and extensive media contacts (Russell, Voas, DeJong, & Chaloupka, 1995).

Other legislative-media efforts, however, have not met with such resounding success. For instance, in 1991, the Massachusetts legislature was considering the passage of a bill that would have allowed the refusal of an allegedly drunk driver to submit to a breathalyzer test to be used as evidence against him or her in the context of a criminal trial. A particular legislator, however, used parliamentary tactics to prevent the passage of the proposed legislation. MADD used extensive media advocacy efforts to alert the public to the issue. Although the bill was ultimately passed and signed, the actions of MADD led to the public exposure of schisms within the organization, which, in turn, reduced its ultimate effectiveness (DeJong, 1996).

Organizational Characteristics

MADD's success as an organization is notable. By 1986, MADD had an annual budget of $7.9 million, 385 chapters in 49 states, and approximately 600,000 members in 49 states, Canada, New Zealand, and England (Schaet, 1986).

MADD and its various chapters have a structure that is similar to a franchise system. A group of individual must contact MADD about setting up a chapter. There must be a minimum of 10 members. Local chapters can collect membership dues as well as solicit other funding, in order to develop their own financial resources. Chapters choose their own local leaders and are free to develop their own programs. Chapters are free to disagree with the central office on a variety of matters. The more successful chapters seem to develop their own direction, although they may participate with the central office in cooperative activities (Weed, 1991).

MADD's success has been attributed to a number of factors including (1) its well-defined niche developed through a focus on drunk driving to the exclusion of numerous other, more divisive, issues of highway safety (Weed, 1987); (2) its minimalist character, that is, its relatively low initial cost, low maintenance cost, the development of supplemental organizational resources in the environment, and a high degree of adaptability (Halliday, Powers, & Granfors, 1987). Volunteer labor has played a critical role in MADD's success (McCarthy & Wolfson, 1996). The emphasis placed on victim services has enhanced MADD's ability to raise funding (McCarthy & Wolfson, 1996).

DISCUSSION QUESTIONS

1. The anti-drunk driving movement, to a large degree, was not founded on epidemiological findings, but rather on advocacy efforts stemming from personal experience.
   a. To what extent does the relationship between alcohol use and traffic accidents as portrayed by advocacy efforts reflect epidemiological findings related to the relationship, if any, between traffic accidents and alcohol?

b. What are the short- and long-term implications of the formulation of law and policy on the basis of advocacy efforts resting on personal experience, rather than on epidemiological findings?

2. Consider the strategies utilized by MADD to effectuate its goals. To what extent is MADD's success attributable to media advocacy efforts? To legislative advocacy efforts?

REFERENCES

Barr, A. (1996). *Drink: A Social History of America.* New York: Carroll & Graf Publishers, Inc.

Berkow, R. (1992). *The Merck Manual of Diagnosis and Therapy.* Rahway, New Jersey: Merck Research Laboratories.

Bordin, R. (1981). *Women and Temperance: The Quest for Power and Liberty, 1873–1900.* Philadelphia: Temple University Press.

Borkman, T. (1976). Experiential knowledge: A new concept for the analysis of self-help groups. *Social Service Review, 50,* 445–456.

Cellar, D. F., Nelson, Z. C., & Yorke, C. M. (2000). The five-factor model and driving behavior: Personality and involvement in vehicular accidents. *Psychological Reports, 86,* 454–456.

Christie, N. (1986). The ideal victim. In E. A. Fattah (ed.), *From Crime Policy to Victim Policy* (pp. 17–30). New York: St. Martin's Press.

DeJong, W. (1996). MADD Massachusetts versus Senator Burke: A media advocacy case study. *Health Education Quarterly, 23,* 318–329.

Duster, T. (1970). *The Legislation of Morality.* New York: Free Press.

Freund, C. P. (1985). Less filling, tastes great. *City Paper* (Washington, D.C.), Feb. 1–8.

Halliday, T., Powell, M. J., & Granfors, M. W. (1987). Minimalist organizations: Vital events in state bar associations, 1870–1930. *American Sociological Review, 52,* 456–471.

Harr, J. (1995). *Civil Action.* New York: Vintage Books.

Hingson, R., Howland, J., Morelock, S., & Heeren, T. (1988). Legal interventions to reduce drunken driving and related fatalities among youthful drivers. *Alcohol Drugs Driving, 4,* 87–98.

Irwin, M., Fitzgerald, C., & Berg, W. P. (2000). The effect of the intensity of wireless telephone conversations on reaction time in a braking response. *Perceptual and Motor Skills, 90,* 1130–1134.

Knoke, D. (1988). Incentives in collective action organizations. *American Sociological Review, 53,* 311–329.

Lemmens, P. H., Vaeth, P. A. C., & Greenfield, T. K. (1999). Coverage of beverage alcohol issues in the print media in the United States, 1985–1991. *American Journal of Public Health, 89,* 1555–1560.

Lerner, M. J. (1980). *The Belief in a Just World: A Fundamental Delusion.* New York: Plenum Press.

Li, H. Z., & Rosenblood, L. (1994). Exploring factors influencing alcohol consumption patterns among Chinese and Caucasians. *Journal of Studies on Alcohol, 55,* 427–433.

Linsky, A. (1971). Theories of behavior and the image of the alcoholic in popular magazines, 1900–1960. *Public Opinion Quarterly, 34,* 573–581.

MADD. (1984). *Organizing a Chapter.* Hurst, Texas: Mothers Against Drunk Driving.

MADD. (1985). *Victim Information Pamphlet.* Hurst, Texas: Mothers Against Drunk Driving.

Mancall, P. C. (195). *Deadly Medicine: Indians and Alcohol in Early America.* Ithaca, New York: Cornell University Press.

Marshall, M., & Oleson, A. (1994). In the pink: MADD and public health policy in the 1990s. *Journal of Public Health Policy, Spring,* 54–68.

McCarthy, J. D., & Wolfson, M. (1996). Resource mobilization by local social movement organizations: Agency, strategy and organization in the movement against drinking and driving. *American Sociological Review, 61,* 1070–1088.

McKenzie, D., & Giesbrecht, N. (1981). Changing perceptions of the consequences of alcohol consumption in Ontario, 1950–1981. *Contemporary Drug Problems, 10*, 215–242.

Murdock, C. G. (1998). *Domesticating Drink: Women, Men, and Alcohol in America, 1870–1940*. Baltimore: Johns Hopkins University Press.

Nawakami, N., Haratani, T., Hemmi, T., & Araki, S. (1992). Prevalence and demographic correlates of alcoholl-related problems in Japanese employees. *Social Psychiatry and Psychiatric Epidemiology, 27*, 198–202.

Reinarman, C. (1988). The social construction of an alcohol problem: The case of Mothers Against Drunk Drivers and social control in the 1980s. *Theory and Society, 17*, 91–120.

Rock, P. (1973). *Deviant Behavior*. London: Hutchison University Library.

Russell, A., Voas, R. B., DeJong, W., Chaloupka, M. (1995). MADD rates the states: A media advocacy event to advance the agenda against alcohol-impaired driving. *Public Health Reports, 110*, 240–245.

Sadoff, M. (1989). Voiceover. *MAADVOCATE*, 2, 4.

Schaet, D. (1986). *MADD Newsletter*, February.

Simos, B. G. (1979). *A Time to Grieve: Loss as a Universal Human Experience*. New York: Family Service Association of America.

Weed, F. J. (1987). Grass-roots activism and the drunk driving issue: A survey of MADD chapters. *Law and Policy, 9*, 259–278.

Weed, F. J. (1991). Organizational mortality in the anti-drunk-driving movement: Failure among local MADD chapters. *Social Forces, 69*, 851–868.

Weed, F. J. (1990). The victim-activist role in the anti-drunk driving movement. *The Sociological Quarterly, 31*, 459–473.

Wortman, C.B. (1983). Coping with victimization: Conclusions and implications for future research. *Journal of Social Issues, 39*, 195–221.

CHAPTER 9

Case Study Six

Needle Exchange Programs

This cry cannot be muted. Those who suffer unjustly have a right to complain and protest. Their cry expresses both their bewilderment and their faith (Gutierrez, 1987: 101).

DEFINING NEEDLE EXCHANGE

Needle Exchange and Harm Reduction

The primary purpose of needle exchange programs has been explained as the reduction of "the spread of infectious disease (including hepatitis and HIV) among IVDUs [intravenous drug users], their sexual partners, and the public at large" Ferrini, 2000: 173). Needle exchange is one facet of what is known as harm reduction, one of several approaches to drug services. Harm reduction posits that the major problem is not the drug use itself, but rather its life effects. Accordingly, the management of the services must be tailored to the individual and his or her needs, with an emphasis on long term support in order to maintain stability (Sorge, 1991).

Unlike many approaches to drug services, the ultimate goal of harm reduction is not abstinence. Rather, abstinence is seen as one end of a continuum of behaviors (Sorge, 1991). Intermediate steps between an individual's current drug use and complete abstinence are seen as valuable; relapse is not seen as failure, but instead as a common occurrence in the attempt to reach abstinence. Accordingly, there exists a range of behavior between abuse and abstinence, which can be ranked in terms of their safety.

The focus of harm reduction is on the social and environmental aspects of drug taking. Interventions are implemented based upon information relating to the experience, organization, and control of individual's drug use. Most drug users are not able to leave their drug-using circumstances even after treatment, so harm reduction focuses on helping injection drug users make positive use of their communities and social contexts in order to survive. This approach recognizes that injection drug users have the ability to make choices, the ability to modify or cease specified behaviors, and the ability to protect themselves (Sorge, 1991). To a great extent, harm reduction programs have focused on individuals who have no contact with more formal systems of providing care, such as social services and drug treatment.

Needle Exchange Programs: History and Examples

Some may question the existence of a link between the development of needle exchange programs and epidemiology. As one researcher has noted,

> With regard to HIV policy, epidemiology has played an important role in the development of needle exchange programs. Early studies showed that injection drug users were an important risk group for HIV (and other bloodborne) infection, that multiple reuse of contaminated syringes was the primary mode of transmission between injection drug users, that injection drug users were an important source of infection to heterosexual women and children, and that a primary reason for the multiple reuse of syringes was the lack of legal access to sterile syringes due to syringe prescription and paraphernalia laws ... .
>
> Epidemiology has provided information critical to forming an empirical basis for a public health program. Epidemiology has a responsibility not only in conducting the studies but—when inconsistencies arise as new studies become available—in performing the investigations necessary to clarify the discrepancies and to guide policymakers and practitioners alike ... In the broader scheme, public health exists in a political context, and epidemiology, as a basic science of public health, is essential for working within the political context. (Vlahov, 2000: 1390–1391).

There are currently at least 40 needle exchange programs (NEPs) in existence (Lloyd, O'Shea, and the Injection Drug Use Study Group, 1994). The first such program to operate in the United States with community support was started in Tacoma, Washington, in 1988. Many such programs allow injecting drug users to exchange needles or syringes for new ones (Lurie et al., 1993b). Several studies have found that the return rate for the injection equipment is high (Buning, 1991; Lurie et al., 1993a).

Approximately one-half of the needle exchange programs in the United States are legal. However, unlike many needle exchange programs in Europe and Canada, the federal government has not been involved in their establishment. Lurie and Chen (1993) have categorized needle exchange programs by level of legality. NEPs are (1) legal, if they are located in a state that does not have a prescription law or that exempts NEPs from such a law; (2) illegal but tolerated, if they are located in a state that has a prescription law but a locally-elected body has voted to support or approve the NEPs; and (3) illegal-underground, if the NEPs have not received such support.

Four administrative models have been identified for NEPs: (1) activist-run NEPs, (2) community-based organizations without government support, (3) community-based organizations with government support, and (4) state- and local-sponsored NEPs (Lurie and Chen, 1993). One program utilizing each of these models is reviewed below.

An Activist-Run NEP: San Diego, California

It has been estimated that there are between 7,153 and 23,381 injection drug users in San Diego County (Green, 1993). This is not surprising, in view of San Diego's ranking by the Department of Justice as third nationally for the percentage of arrestees who test positive for heroin, first or second for rates of polydrug use, and first in rates of methamphetamine use (SANDAG, 1994). Additionally, San Diego has been designated as a "high intensity drug-trafficking area" (San Diego Department of Health Services, Office of AIDS Coordination, 1994). In addition, due to the proximity of San Diego to Mexico, where syringes and needles and many injectable drugs are available without prescription over the counter, individuals may purchase vitamins, steroids, hormones, and antibiotics, as well as injection equipment in Mexico and then self-medicate at home, possibly sharing their equipment with others. It has been estimated that between 5 and 10 percent of injection drug users who are in treatment programs test positive for HIV (San Diego Department of Health Services, Office of AIDS Coordination, 1993).

The San Diego needle exchange program was begun on January 8, 1992 by volunteers with ACT-UP. (See discussion below). The program, which is illegal, has been able to survive through the donations and support of volunteers and private grants. In July 1994, the program became known as the San Diego Clean Needle Exchange (Lloyd, O'Shea, and the Injection Drug Use Study Group, 1994).

The program exchanged more than 32,000 needles and served more than 5,000 clients between January 8, 1992 and February 1994, when one of the exchange members was arrested. Since that arrest, exchanges have been effectuated on an in-home basis. In addition to providing clean needles and syringes, the visits provide an opportunity to distribute educational materials relating to HIV prevention and testing, condoms, bleach, cotton, alcohol wipes, and relevant community resources. The program exchanges an average of 600 to 1,000 syringes per week. The clients of the program range in age from 20 to 60. Approximately 60% are African American, 30% are Latino, and 10% identify as Caucasian.

A CBO Without Government Support: Bronx-Harlem

Approximately 200,000 injections drug users live in the New York City metropolitan area. More than half of the 8,447 narcotic-related deaths that occurred between 1978 and 1986 were due to AIDS. HIV prevalence among New York's injection drug users has been estimated to lie between 55% and 60%, making it the highest in the United States. Of New York's approximately 39,000 to 43,000 available drug treatment slots, only 20% are devoted to non-heroin users and to programs other than methadone maintenance. The South Bronx, which contains 38% of the borough's population, is home to 66% of the IDU population. Approximately 1% of the population of the Bronx is infected with HIV (Lurie et al., 1993a).

New York City's Health Commissioner approved the establishment of a needle exchange program in January 1988. Ultimately, the program was limited to one site, located near a police station, as the result of community opposition. The mayor ultimately closed the program in February 1990 due to its limited usefulness, which resulted from various location and administrative constraints (Lurie et al., 1993a).

In February 1990, the AIDS Coalition to Unleash Power (ACT-UP) started a needle exchange program to protest the closure of the city's program. The underground exchange evolved into two sites, one in Bronx-Harlem and one on the Lower East Side. The mayor announced formal support for these programs following the review of the research conducted to evaluate the New Haven needle exchange program. (See below.) The New York State Health Commissioner on July 19, 1992 granted waivers of state prescription laws to the two programs, thereby allowing the possession of a syringe without a prescription (Lurie et al., 1993a).

The Bronx-Harlem NEP is located in a traditional drug trafficking and sales area. The needle exchange program has received funding from the American Foundation for AIDS Research (AmFAR), which is administered by the state of New York.

A CBO With Government Support: New Haven, Connecticut

There are approximately 1,910 to 2,660 injection drug users in New Haven. The rate of AIDS among the IDUs is 13 per 100,000; among gay and bisexual IDUs, it is 0.04 per 100,000 (Lurie et al., 1993a). The needle exchange program in New Haven began in 1986 through a group that became known as the National AIDS brigade (NAB). The needle exchange program that was developed through the New Haven Department of Health officially began in November 1990 as an outgrowth of the Mayor's Task Force on AIDS. The General Assembly of Connecticut gave legal sanction to the program in 1990 and allocated $25,000 to it. By 1992, it had received more than $200,000. Later, legislation was adopted in 1992 to permit pharmacists to sell up to 10 syringes at a time over the counter. The legislation also authorized the expansion of the needle exchange program to other cities.

A variety of services are offered through the program, including syringes, cotton, cookers, bleach, condoms, sterile rinse water, health education materials, HIV testing, and referrals for tuberculosis, testing for sexually transmitted diseases including HIV, primary health care, social services, counseling, and housing. Syringes distributed through the program are marked for tracking. The overall return rate for syringes given out by the exchange program is 68%. It has been estimated that the New Haven needle exchange program reaches from 48% to 68% of injection drug users in New Haven. Approximately 41% of the clients are African American, 34% Caucasian, and 25% Latino, based on self-reported characteristics.

It has been found that the needle exchange program has decreased the percentage of syringes testing positive for HIV and has served as an entry point for drug treatment services (Heimer, Kaplan, Khoshnood, Jariwala, and Cadman, 1993). Approximately one-third of the clients had requested assistance in entering drug treatment programs; the exchange program was able to place over one-half of those requesting such services.

A Locally-Sponsored Program: Seattle, Washington

Approximately 12,000 injection drug users live in the Seattle-King County area. Of these, it is estimated that as many as 5% are HIV-infected (Lurie et al., 1993a).

Needle exchange began in Seattle in March 1989 through the efforts of ACT-UP. These efforts were supported by the Public Health Department, which provided outreach workers with materials on drug abuse treatment and HIV prevention to be distributed to their clients. The State Pharmacists' Association provided the first supply of syringes. The Public Health Department took over the exchange program the following month, recognizing the program as a vehicle for reaching injection drug users (Lurie et al., 1993a).

The needle exchange program operates at several sites. By 1990, approximately 2,342 individuals had contact with the needle exchange program annually. The Public Health Department distributed over one million syringes between the second quarter of 1989 and the first quarter of 1993. Additionally, the NEP has been successful in assisting its clients obtain drug treatment services.

THE BASIS FOR NEEDLE EXCHANGE

One must necessarily question why needle exchange programs have been developed, why traditional, abstinence-based programs are not sufficient. The answer to this question is multi-faceted and requires and understanding of injection drug use and its consequences.

Injection Drug Use: Behavior and Biology

Injection behavior varies according to the drug that is being used. For instance, some drugs, such as cocaine, require more frequent injections to maintain a feeling of euphoria and to prevent the onset of withdrawal symptoms (Des Jarlais et al., 1988). Behavior following injection similarly varies with the drug being used, due to differences in the pharmacological properties across various drugs. For instance, users of cocaine may become stimulated, excited, talkative, and aggressive, while heroin users may "nod out," that is, display depressed speech and reduced motor activity (Chiasson, Stoneburner, Telzak, Hildebrandt, Schultz, & Jaffe, 1989). Individuals who inject more frequently in an effort to avoid or reduce the acute symptoms of withdrawal may potentially be more likely to abandon safer needle hygiene practices (Chiasson, Stoneburner, Telzak, Hildebrandt, Schultz, & Jaffe, 1989).

Drugs that are ingested orally are dissolved, digested, and dispersed, and are then absorbed from the gastrointestinal tract into the bloodstream, by which they reach the brain. The majority of drugs are metabolized in the liver, where they are made less toxic. Because of this, drugs that are taken orally may not take effect for an hour or so. Drugs that are injected, however, are injected directly into the bloodstream, into a muscle mass, or under the upper layers of skin. Because the drug does not have to be absorbed, the effect is much more rapid. Consequently, less drug will be required to get the same degree of euphoria, or "high." The intravenous injection of a drug into a vein will reach the brain in approximately 10 to 20 seconds, in a small compact ball, known as a bolus. The effect will be both very rapid and very intense, a "rush." The "rush" is not dependent on an individual's past use, as it is entirely a physiological response to the drug. The adverse effects of the drug on the body potentially are more severe when it is injected, as compared to ingested, because its effects are not mediated by the metabolic action of the liver (Stryker, 1989).

Even irritating substances can be injected because the walls of blood vessels are relatively insensitive. However, repeated injections in the same vein wall causes the wall to lose its strength and elasticity and, eventually, to collapse. When this happens, another vein must be found for the injections (Ray & Ksir, 1990). Subcutaneous injection of irritating substances may result in "skin popping," whereby the skin around the injection site dies and sheds (Ray & Ksir, 190).

The Impact of Substance Use and Abuse

Drug use has a huge impact on both morbidity and mortality. Almost one-half million people die each year in the United States due to substance abuse (Robert Wood Johnson Foundation, 1993). Almost 40 percent of the drug-related deaths are among adults between the ages of 30 and 39. More than one-third of new AIDS cases occur among individuals injecting drugs or among those who have sexual contact with them (Robert Wood Johnson Foundation, 1993).

The rates of frequent heavy drug use, measured on a daily or weekly basis, remain unchanged, although there is some indication that overall consumption has declined among some portions of the population. The National Institute on Drug Abuse has estimated that at least 74 million Americans have used at least one illegal substance in their lifetime. It has also been estimated that 1.3 million Americans inject drugs yearly; approximately one-half of these individuals use heroin and the remaining half use cocaine and methamphetamines (*AIDS Alert*, 1994).

The impact of substance use is also felt by children born to mothers who have ingested substances during their pregnancies (Faden & Graubard, 2000; Pollard, 2000). Prolonged use of illicit drugs can result in the impairment of various organ systems. The use of alcohol and illicit drugs has been found to be associated with an increased risk of HIV transmission because individuals who have used drugs or alcohol are more likely to engage in unsafe sexual practices (Weinhardt, Carey, Carey, Maisto, & Gordon, 2001; Stein, Hanna, Natarajan, Clarke, Marisi, Sobota, & Rich, 2000).

Substance use results in costs to employers, through missed time and days at work, related injuries, and treatment (Robert Wood Johnson Foundation, 1993). The economic costs to the United States have been estimated to be upwards of $238 billion, representing one-quarter of the nation's healthcare costs (Robert Wood Johnson Foundation, 1993).

HIV, AIDS, and Injection Drug Use

HIV/AIDS: The Very Basics

Human immunodeficiency virus (HIV) is believed to be the causative agent of acquired immune deficiency syndrome (AIDS). AIDS, which was first recognized as a distinct syndrome in 1981, is a life-threatening clinical condition.

Several weeks or months after infection with HIV, some individuals develop a mononucleosis-type illness that lasts approximately one to two weeks. Often, individuals will then remain symptom-free for many years prior to the development of clinical manifestations. These initial symptoms may include lymphadenopathy, anorexia, chronic diarrhea, weight loss, fever, and fatigue. As the disease progresses, individuals may develop numerous opportunistic infections, such as Kaposi's sarcoma, toxoplasmosis of the CNS, and extrapulmonary tuberculosis. HIV can be transmitted through contact with body fluids, such as semen, vaginal fluid, and blood. Consequently, sexual contact, the use of contaminated injection equipment, the receipt of contaminated blood or blood products constitute the primary modes of transmission, and vertical transmission from mother to child remain the primary modes of transmission (Osmond, 1999).

Injection Drug Use and HIV Transmission

As of June 1996, CDC had found that 46% of women with AIDS had injected drugs, 18% of women with AIDS were sexual partners of IDUs, and 54% of infants diagnosed with AIDS were born to mothers who injected drugs or who were sexual partners of IDUs. Further, 36% of all African Americans with AIDS and 37% of all Hispanics with AIDS were IDUs (CDC, 1996).

HIV can be transmitted from an HIV-infected injection drug user to another injection drug user through the shared use of contaminated equipment. Blood may be introduced into the needle and syringe through "registering" or through "booting." "Registering" refers to a practice whereby blood is drawn into the syringe to ensure that the needle is actually in a vein so that the drug can be injected intravenously. "Booting" means that the blood is drawn back into the syringe several times in order to utilize any residual drug that may remain in the syringe. Even a miniscule amount of HIV-infected blood in the syringe may result in the transmission of the infection from one user to the next (Centers for Disease Control, 1993). The probability of HIV infection for each injection with a contaminated syringe has been estimated to be .0067, which is higher than the probability of HIV infection from a needlestick injury and is three times higher than the probability of infection by an infected man of an uninfected woman as a result of vaginal sex (Kaplan & Heimer, 1992a, 1992b). Transmission may result from the shared use of other contaminated injection equipment, such as cookers (bottle caps or spoons used to heat and dissolve drugs in water), cotton (used as a filter to draw drugs into a syringe), and Q-tips (also used as filters to draw drugs into the syringe).

Other practices may also increase the risk of transmission: squirting the drug solution from a contaminated syringe into the drug mixing cooker or spoon and then drawing it into another syringe; using the plunger from a previously used, contaminated syringe to mix the drug with the water; "beating" used cotton to obtain any drug that remains in the cotton; "kicking out a taste" by putting a part of the drug/water solution from a previously used, contaminated syringe back into the cooker into another syringe so that other IDUs can get some of the same drug; rinsing a previously used, contaminated syringe in water that other IDUs will use to rinse their own syringes or to dissolve drugs; and, finally, drawing up the water for dissolving the drug by using previously used, contaminated syringes (Academy for Educational Development, 1997).

Contrary to the popularized image of injection drug users, many IDUs provide each other with mutual support and share things deemed to be of value, including housing, food, money, and clothing. The sharing of drugs provides some balance to the intense competition in obtaining drugs and injection equipment. IDUs may work in pairs or small groups to provide each other with mutual support, with protection, and to obtain money and drugs (Des Jarlais, Friedman, Sotheran, and Stoneburner, 1988). Friends, lovers, or dealers often supply the drug and the equipment to new injectors. Many IDUs may seek assistance and guidance from those who are more experienced (Grund, Blanken, Adriaans, Kaplan, Barendregt, & Meeeuwsen, 1992). An IDU's refusal to share with someone with whom he or she has previously shared is taken to be a sign of suspicion or mistrust (Stryker, 1989). Various other factors combine to strengthen the bonds between IDUs, including their ostracism and marginalization and economic deprivation, (Feldman and Biernacki, 1988).

Hepatitis and Injection Drug Use

Between 50% and 80% of those who use injection drugs have serologic evidence of infection with hepatitis B, either currently or in the past.

Between 50% and 85% show evidence of infection with hepatitis C (Gerfein, Vlahov, Galai et al., 1996; Kaplan and Heimer, 1992b; Centers for Disease Control, 1990). Testing positive for HCV antibody has been found to be associated with having injected for a longer period of time (Diaz, Des Jarlais, Vlahov, Perlis, Edwards, Friedman, Rickwell, Hover, Williams, and Monterroso, 2001; Lorvick, Kral, Seal, Gee, and Edlin, 2001), with sharing cotton (Diaz, Des Jarlais, Vlahov, Perlis, Edwards, Friedman, Rickwell, Hover, Williams, & Monterroso, 2001), with sharing cookers (Hagan, Thiede, Weiss, Hopkins, Duchin, and Alexander, 2001), with being in drug treatment, and with being older than 24 years (Diaz, Des Jarlais, Vlahov, Perlis, Edwards, Friedman, Rickwell, Hover, Williams, & Monterroso, 2001). It has been estimated that approximately 20% to 30% of those infected with hepatitis C will progress to cirrhosis, hepatocellular carcinoma, or both within 2 to 3 decades (National Institutes of Health, 1997).

Research has consistently demonstrated that needle exchange programs may reduce the rate of hepatitis transmission by one-third or more (Hagan, Des Jarlais, Friedman, Purchase, & Alter, 1995; Hagan, Reid, Des Jarlais et al., 1991; Taylor, 1991).

PUBLIC POLICY AND NEEDLE EXCHANGE

Public attitudes towards illicit drug use has varied over time. Heroin became more widely used during the 1950s and 1960s; the use of drugs, in general, increased among the general public in the 1970s, ultimately peaking in the 1970s for most drugs. Illicit drug use appeared to decrease in most segments of the U.S. population during the 1980s and the 1990s (Robert Wood Johnson Foundation, 1993).

However, heroin regained popularity in the late 1970s, following a period of relative scarcity. Individuals with a history of heroin use often preferred to inject, rather than smoke, cocaine, and combined heroin and cocaine into a "speedball." Unlike heroin, however, treatment for such drug use was relatively unavailable (Des Jarlais, Friedman, Sotheran, & Stoneburner, 1988).

Research examining the effects of needle exchange has consistently found that the provision of clean needles and syringes is an effective strategy for reducing the risk of HIV transmission among those who inject drugs as well as their sexual contacts (Kirp & Bayer, 1992) and that needle exchange does not result in an increase in the number or distribution of discarded needles (Doherty, Junge, Rathouz, Garfein, Riley, & Vlahov, 2000). Exchange programs have been found to be associated with less lending of equipment (Watters, Estilio, Clark, & Lorvick, 1994; Schwartz, 1993), decreased frequency of injection drug use (Watters, Estilio, Clark, & Lorvick, 1994; Guydish, Bucardo, Young et al., 1993), and increased referrals for drug treatment and social services (Kaplan & Heimer, 1992b; Carvell and Hart, 1990). Most recently, the establishment of needle exchange programs has received support from the American College of Preventive Medicine, which stated:

> Needle-exchange programs should be implemented and expanded in areas with high rates of intravenous drug abuse. Although data are still preliminary, they support a public health intervention that is inexpensive (especially in comparison with the societal costs of treating those with HIV) and likely to reduce the transmission of fatal bloodborne infection among drug users, their sexual partners, and their children. These programs should include education and counseling, referral for drug and medical treatment, HIV and hepatitis testing, and condom distribution. Needle exchange programs should be voluntary, anonymous, and accessible and should strive to recapture all needles distributed to reduce the risk to the public. All needle-exchange programs should have an ongoing evaluative component to assess efficacy and need ... Needle-exchange programs should not substitute for a comprehensive approach to drug treatment and prevention (Ferrini, 2000: 174–175).

However, despite such recommendations and support and the relative unavailability of drug treatment and the demonstrated effectiveness of needle exchange programs,

> mainstream American politicians still mostly dismiss the idea of making clean syringes available to drug users as misguided and dangerous—as "sending the wrong message" in the midst of the official war on drugs. American law reflects this hostility. A doctor's prescription is necessary to buy needles in many states, including those where drug injection is most common. And in almost every state, needle possession for the purpose of illegal drug use is outlawed by drug paraphernalia statutes. The United States unintentionally has become the control group in an international investigation of the effectiveness of needle exchange (Kirp & Bayer, 1993: 78–79).

Data suggest that this "control group" is experiencing an increased risk of disease. A cross-sectional analysis of the relationship between laws prohibiting over-the-counter sale of syringes and the prevalence of HIV among injection drug users found that laws restricting syringe access are associated with HIV transmission and are not associated with lower population proportions of injection drug users (Friedman, Perlis, & Des Jarlais, 2001).

The policy against needle exchange reflects the federal government's policy of "zero tolerance" towards drug use and its general hostility towards injection drug users. "Drug Czar" Robert Martinez had denounced the provision of clean needles and syringes to injection drug users as "a retreat in the war against drugs" (Executive Office of the President, 1992), while the former police chief of Los Angeles, Daryl Gates, announced that "We should take users and shoot them" (Kleiman, 1992). Martinez has justified his stance by noting

> Our gains against drug use have been hard-won, and this is no time to jeopardize them by instituting needle exchange programs. Despite all the arguments made by proponents of needle exchange, there is no getting around the fact that distributing needles facilitates drug use and undercuts the credibility of society's message that using drugs is illegal and morally wrong (Martinez, 1992).

The policy, however, is congruent with the objections that have been voiced by many African American politicians and leaders. Some

African-Americans may see needle exchange as yet another social experiment visited upon them by a white establishment unconcerned with their well-being. Others see needle exchange as yet another form of genocide (Kirp & Bayer, 1993).

COMMUNITY ACTIVISM AND THE ESTABLISHMENT OF NEEDLE EXCHANGE PROGRAMS

To a great extent, the establishment of needle exchange programs, such as those described above, has derived from the efforts of charismatic individuals. Some of these individuals have relied on confrontational strategies, while others have employed a social work-activism model (Kirp & Bayer, 1993). Jon Parker, a former injection drug user, was one of the first, if not the first person to distribute injection equipment publicly in the United States. He has been arrested in at least eight states and has challenged the law in all of the states where the purchase of needles and syringes without a prescription remains illegal (Lane, 1993). Similarly, Dave Purchase organized the needle exchange in Tacoma, Washington after notifying potentially interested officials that he would begin such a program. The volunteer staff of Prevention Point in San Francisco opened the exchange program on November 2, 1988, the Day of the Dead (El Dia de los Muertos); many were activists who launched the program specifically as an act of civil disobedience. One staff member explained:

> The clients ... observed us taking a risk of arrest, which in the [drug-using] community has tremendous social meaning. And we did it night after night. The night [in October 1989] after the earthquake in San Francisco, we were there with flashlights .... So, when they saw this behavior with people they perceived [as having] something to lose, they perceived that we were more like them. They were carrying heroin in their pockets, they were illegal; we were carrying needles, we were illegal. And there was a reduction in the social distance (Patricia Case, 1992, quoted in Lane, 1993: 5).

Support has been received from some private foundations. What has been especially critical has been the endorsement of the concept and the programs by public health leaders.

DISCUSSION QUESTIONS

1. Compare and contrast MADD's efforts to reduce drunk driving with the efforts of advocates to establish needle exchange programs.
   a. Comparing the two movements, to what extent did each utilize epidemiological data as an element of its strategy?
   b. What strategies were utilized by each movement and to what extent were they or were they not successful?
2. Researchers analyzing the success of MADD have emphasized its self-portrayal as the voice of the victim.

a. To what extent has the needle exchange movement reflected the voice of the victim?
b. How is the nature of that victim, if any, the same or different as compared to the victim portrayed through MADD's efforts?
c. How have those similarities and differences, if any, affected the ability of advocacy efforts for needle exchange to meet their goals?

REFERENCES

Academy for Educational Development. (1997). *HIV Prevention Among Drug Users: A Resource Book for Community Planners and Program Managers.* June.

AIDS Alert. (1994). Common sense about AIDS avoiding alcohol, illegal drugs may reduce HIV exposure risk. *AIDS Alert, 9,* insert.

Buning, E. C. (1991). Effects of Amsterdam needle and syringe exchange. *International Journal of the Addictions, 26,* 1303–1311.

Carvell, A. M., & Hart, G. J. (1990). Help-seeking and referrals in a needle exchange: A comprehensive service to injecting drug users. *British Journal of Addiction, 85,* 235–240.

Centers for Disease Control. (1993). *Facts About Drug Use and HIV/AIDS.* Atlanta, Georgia: Centers for Disease Control and Prevention.

Centers for Disease Control. (1996). *HIVAIDS Surveillance Report.* Atlanta, Georgia: Centers for Disease Control and Prevention.

Centers for Disease Control. (1990). Protection against viral hepatitis: Recommendations of the Immunization Practices Advisory Committee. *Morbidity and Mortality Weekly Report, 39,* 1–26.

Chiasson, M. A., Stoneburner, R. L., Telzak, E., Hildebrandt, D., Schultz, S., & Jaffe, H. (1989). Risk factors for HIV-1 infection in STD clinic patients: Evidence for crack-related heterosexual transmission. Presented at the Fifth International AIDS Conference, Montreal, June.

Des Jarlais, D. C., Friedman, S. R., Sotheran, J. L., & Stoneburner, R. (1988). The sharing of drug injection equipment and the AIDS epidemic in New York City. In *Needle Sharing Among Intravenous Drug Abusers: National and International Perspectives.* Institute on Drug Abuse Research Monograph Series.

Diaz, T., Des Jarlais, D. C., Vlahov, D., Perlis, T. E., Edwards, V., Friedman, S. R., Rockwell, R., Hoover, D., Williams, I. T., & Monteroso, E. R. (2002). Factors associated with prevalent hepatitis C: Differences among young adult injection drug users in lower and upper Manhattan, New York City. *American Journal of Public Health, 91,* 23–30.

Doherty, M. C., Junge, B., Rathouz, P., Garfein, R. S., Riley, E., & Vlahov, D. (2000). The effect of a needle exchange program on numbers of discarded needles: A 2-year follow-up. *American Journal of Public Health, 90,* 936–939.

Executive Office of the President, Office of National Drug Control Policy. (1992). Needle exchange programs: Are they effective? *ONDCP Bulletin, 7,* 1–7.

Faden, V. B., & Graubard, B. I. (2001). Maternal substance use during pregnancy and developmental outcome at age three. *Journal of Substance Abuse, 12,* 329–340.

Feldman, H. W., & Biernacki, P. (1988). The ethnography of needle sharing among intravenous drug users and implications for public policies and intervention strategies. In *Needle Sharing Among Intravenous Drug Abusers: National and International Perspectives* (pp. 28–39). National Institute on Drug Abuse, Research Monograph Series.

Ferrini, R. (2000). American College of Preventive Medicine public policy on needle-exchange programs to reduce drug-associated morbidity and mortality. *American Journal of Preventive Medicine, 18,* 173–175.

Friedman, S. R., Perlis, T., & Des Jarlais, D. C. (2001). Laws prohibiting over-the-counter syringe sales to injection drug users: Relations to population density, HIV prevalence, and HIV incidence. *American Journal of Public Health, 91,* 791–793.

Garfein, R. S., Vlahov, D., Galai, N. et al. (1996). Viral infections in short-term injection drug users: The prevalence of hepatitis C, hepatitis B, human immunodeficiency, and human T-lymphotropic viruses. *American Journal of Public Health, 86*, 655–661.

Green, J. O. (1993). *Estimates of Drug Users in San Diego County, 1990–1992, Final Report.* Prepared for the San Diego County Department of Health Services, Alcohol and Drug Services, San Diego, California. June.

Grund, J. P. C., Blanken, P., Adriaans, N. F. P., Kaplan, C. D., Barendregt, C., & Meeeuwsen, M. (1992). Reaching the unreached: Targeting hidden IDU populations with clean needles via known user groups. *Journal of Psychoactive Drugs, 24*, 41–47.

Guydish, J., Bucardo, J., Young, M. et al. (193). Evaluating needle exchange: Are there negative effects? *AIDS, 7*, 871–876.

Hagan, H., Des Jarlais, D. C., Friedman, S. R., Purchase, D., & Alter, M. J. (1995). Reduced risk of hepatitis B and hepatitis C among injection drug users in the Tacoma Syringe Exchange Program. American Journal of Public Health, *85*, 1531–1537.

Hagan, H., Reid, T., Des Jarlais, D. C. et al. (1991). The incidence of HBV infection and syringe exchange programs. *Journal of the American Medical Association, 266*, 1646–1647.

Hagan, H., Thiede, H., Weiss, N. S.,, Hopkins, S. G., Duchin, J. S., & Alexander, E. R. (2001). Sharing of drug preparation equipment as a risk factor for hepatitis C. *American Journal of Public Health, 91*, 42–46.

Heimer, R., Kaplan, E. H., Khoshnood, K., Jariwala, B., & Cadman, E. C. (1993). Needle exchange decreases the prevalence of HIV-1 proviral DNA in returned syringes in New Haven, Connecticut. *American Journal of Medicine, 95*, 214–220.

Kaplan, E. H., & Heimer, R. (1992a). A model-based estimate of HIV infectivity via needle sharing. *Journal of Acquired Immune Deficiency Syndromes, 5*, 116–118.

Kaplan, E. H., & Heimer, R. (1992b). HIV prevalence among intravenous drug users: Model-based estimates from New Haven's legal needle exchange. *Journal of Acquired Immune Deficiency Syndromes, 5*, 163–169.

Kirp, D. L., & Bayer, R., eds. (1992). *AIDS in the Industrialized Democracies: Passions, Politics, and Policies.* New Brunswick, New Jersey: Rutgers University Press.

Kirp, D.L., Bayer, R. (1993). The politics. In J. Stryker, & M.D. Smith (Eds.)., *Dimensions of HIV Prevention: Needle Exchange* (pp. 77–97). Menlo Park, California: Henry J. Kaiser Family Foundation.

Kleiman, M. (1992). *Against Excess: Drug Policy for Results.* New York: Basic Books.

Lane, S. D. (1993). A brief history. In In J. Stryker, & M. D. Smith (Eds.)., *Dimensions of HIV Prevention: Needle Exchange* (pp. 1–9). Menlo Park, California: Henry J. Kaiser Family Foundation.

Lloyd, L. S., & O'Shea, D. J., Injection Drug Use Study Group. (1994). *Injection Drug Use in San Diego County: A Needs Assessment.* San Diego: Alliance Healthcare Foundation, October.

Lorvick, J., Kral, A. H., Seal, K., Gee, L., & Edlin, B. R. (2001). Prevalence and duration of hepatitis C among injection drug users in San Francisco, California. *American Journal of Public Health, 91*, 46–47.

Lurie P, Reingold, A. L., Lee, P. R., Bowser, B., Chen, D., Foley, J., Guydish, J., Kahn, J. G., Lane, S., & Sorenson, J. (1993a). *The Public Health Impact of Needle Exchange Programs in the United States and Abroad, Volumes 1 and 2.* Prepared for the Centers for Disease Control and Prevention (CDC) by the School of Public Health, University of California, Berkeley, and the Institute for Health Policy Studies, University of California, San Francisco. Atlanta, Georgia: CDC.

Lurie. P., Reingold, A. L., Lee, P. R., Bowser, B., Chen, D., Foley, J., Guydish, J., Kahn, J. G., Lane, S., & Sorenson, J. (1993). *The Public Health Impact of Needle Exchange Programs in the United States and Abroad: Summary, Conclusions, and Recommendations.* Prepared for the Centers for Disease Control and Prevention (CDC) by the School of Public Health, University of California, Berkeley, and the Institute for Health Policy Studies, University of California, San Francisco. Atlanta, Georgia: CDC.

Lurie, P., & Chen, D. (1993). A review of programs in North America. In *Dimensions of HIV Prevention: Needle Exchange* (J. Stryker, M. B. Smith, eds.). Menlo Park, California: The Kaiser Forums, The Kaiser Family Foundation.

Martinez, R. (1992). Executive Office of the President, Office of National Drug Control Policy. Letter to ONDCP Bulletin, 7, Needle exchange programs, Are they effective? July.

National Institutes of Health. (1997). *Management of Hepatitis C. Consensus Statement 15.* Washington, D.C.: National Institutes of Health.

Osmond, D. H. (1999). Epidemiology of HIV/AIDS in the United States. In P. T. Cohen, M. A. Sande, P. A. Volberding (eds.). *The AIDS Knowledge Base: A Textbook on HIV Disease from the University of California, San Francisco, and the San Francisco General Hospital* (pp. 13–21). Philadelphia: Lippincott Williams & Wilkins.

Pollard, I. (2000). Substance abuse and parenthood: Biological mechanisms-bioethical challenges. *Women and Health, 30,* 1–24.

Ray, O., & Ksir, C. (1990). Drugs, Society, and Human Behavior. St. Louis, Missouri: Times Mirror/ Mosby College Publishing.

Robert Wood Johnson Foundation. (1993). *Substance Abuse: The Nation's Number One Health Problem. Key Indicators for Policy.* Prepared by the Institute for Health Policy, Brandeis University, October.

San Diego Department of Health Services, Office of AIDS Coordination. (1993). *Baseline HIV/AIDS Needs Assessment for San Diego County.* San Diego, California: San Diego County Department of Health Services, Office of AIDS Coordination, September.

San Diego Department of Health Services, Office of AIDS Coordination. (1994). County of San Diego FY 1994 Supplemental Application for Ryan White Comprehensive Emergency Act of 1990 Emergency Relief Grant Program. San Diego, California: San Diego County Department of Health Services, Office of AIDS Coordination, January 11.

Schwartz, R. H. A. (1993). Syringe and needle exchange programs: Part I. *Southern Medical Journal, 86,* 318–322.

Stein, M. D., Hanna, L., Natarajan, R., Clarke, J., Marisi, M., Sobota, M., & Rich, J. (2000). Alcohol use patterns predict high-risk HIV behaviors among active injection drug users. *Journal of Substance Abuse Treatment, 18,* 359–363.

Stryker, J. (1989). IV drug use and AIDS: Public policy and dirty needles. *Journal of Health Politics, Policy and Law, 14,* 719–740.

Taylor, F. (1991). Decline in hepatitis B cases. *American Journal of Public Health, 81,* 221–222.

Watters, J. K., Estilio, M. J., Clark, G. L., & Lorvick, J. Syringe and needle exchange as HIV/AIDS prevention for injection drug users. *Journal of the American Medical Association, 271,* 115–120.

Weinhardt, L. S., Carey, M. P., Carey, K. B., Maisto, S. A., & Gordon, C. M. (2001). The relation of alcohol use to HIV-risk sexual behavior among adults with severe and persistent mental illness. *Journal of Consulting and Clinical Psychology, 69,* 77–84.

CHAPTER 10

Epidemiology, Law, and Social Context

[E]pidemiology is inevitably entangled with society, and it is not feasible or desirable to study the causes of disease in the abstract (Pearce, 1996: 682).

The most significant insight that has emerged from sociological studies of science in the past 15 years is the view that science is *socially constructed* .... "[F]acts" are produced by human agency through the institutions and processes of science, and hence they invariably contain a social component ... Form a sociological viewpoint, scientific claims are never absolutely true but are always *contingent* on such factors as the experimental or interpretive conventions that have been agreed to within relevant scientific communities .... [S]cience as we know it often takes the form of written texts or *inscriptions*,such as a curve on graph paper, a scattering of dots on photographic films, or an x-ray picture that looks like a supermarket bar code .... The inscription is a substitute for reality .... For sociologists of science, deconstruction means nothing more arcane than the pulling apart of socially constructed facts ... (Jasanoff, 1993: 77–82).

A number of studies have linked the history of specific disease episodes—malaria epidemics, outbreaks of typhus, or of smallpox—to wider patterns of political and economic change. Yet in limiting their time frame they have been unable to describe how these linkages have evolved over longer periods of time and how realignments in specific sets of political and economic interests have shaped the longer history of both health and health care (Packard, 1989: 20).

DEVIANCE IN SOCIAL CONTEXT

A 1965 (U.S.) study of public perceptions of deviance revealed wide diversity in what people classify as deviant: homosexuals (49% of responses), drug addicts (47%), alcoholics (46%), prostitutes (27%), murderers (22%), lesbians (13%), the mentally ill (12%), perverts (12%), communists (10%), and atheists (10%), to name a few (Simmons, 1965). Various approaches have been offered in an attempt to better understand the nature of deviance and why there exists both variation and similarity in perceptions of deviance. Orcutt (1983) has suggested that deviance can be viewed alternatively as: (1) behavior that violates social norms or (2) behaviors that are defined as deviant by the social audiences viewing them. The former definition is termed the "normative" perspective, while the latter is termed "relativistic."

The Normative View of Deviance

Norm Violation

The normative view of deviance is premised on the observation that rules or expectations for behavior, *i.e.* norms, are shared by members of a group or society (Orcutt, 1983). Although Edgerton (1985) premises his discussion of deviance on rule violation, the concept of the norm is closely related. Edgerton (1985: 24), for instance, has defined a rule as "a shared understanding of how people ought to behave and what should be done if someone acts in a way that conflicts with that understanding." Edgerton (1985: 24) further explains that

> rules may differ in the extent to which they are known, recognized, accepted as just or proper, and uniformly applied to members of the society. Rules may also vary in the severity of the sanctions that may be incurred by their violations as well as in their consistency of enforcement. They may vary in the degree to which they are internalized, in the mode of their transmission, and in the amount and kind of conformity they receive .... [S]ome rules are relatively explicit while others are implicit, and some clear while others are ambiguous. Some rules contradict others, while others stand for the most part unchallenged.

A "norm" has been variously defined as

> a rule or a standard that governs our conduct in the social situations in which we participate. It is a societal expectation. It is a standard to which we are expected to conform whether we actually do so or not (Bierstedt, 1963: 222).

> shared convictions about the patterns of behavior that are appropriate or inappropriate for the members of a group; what group members agree they can, should, might, must, cannot, should not, ought not, or must not do in any given situation (DeFleur et al., 1977: 620).

> a rule which, over a period of time, proves binding on the overt behavior of each individual in an aggregate of two or more individuals. It is marked by the following characteristics: (1) Being a rule, it has a content known to at least one member of the social aggregate. (2) Being a binding rule, it regulates the behavior of any given individual in the social aggregate by virtue of (a) his having internalized the rule; (b) external sanctions in support of the rule applied to him by one or more other individuals in the social aggregate; (c) external sanctions in support of the rule applied to him by an authority outside the social aggregate; or any combination of these circumstances (Dohrenwend, 1959: 470).

> a rule, standard, or pattern for action ... Social norms are rules for conduct. The norms are the standards by reference to which behavior is judged and approved or disapproved. A norm in this sense is not a statistical average of actual behavior but rather a cultural (shared) definition of desirable behavior (Williams, 1968: 204).

Several elements emerge from these definitions that permit us to better define what is meant by deviance: (1) consensus, both as to what constitutes a rule and a violation of that rule and (2) agreed upon consequences in response to that violation. A review of several definitions of deviance supports this conclusion:

> [D]eviance constitutes only those deviations from norms which are in a disapproved direction and of sufficient degree to exceed the tolerance limits of a social group such that the deviation elicits, or is likely to elicit if detected, a negative sanction (Clinard and Meier, 1979: 14).
>
> Deviance is the name of the conflict game in which individuals or loosely organized small groups with little power are strongly feared by a well-organized, sizable minority or majority who have a large amount of power (Denisoff and McCaghy, 1973: 26).
>
> [D]eviant behavior refers to conduct that departs significantly from the norms set for people in their social statuses (Merton, 1966: 805).
>
> [W]e may define deviant behavior in terms of public consensus. We may define it as *any behavior considered deviant by public consensus that may range from the maximum to the minimum* (Thio, 1978: 23).

However, these definitions leave numerous questions unresolved: Whose consensus? Whose opinion is to be considered in "calculating" consensus and what weight is it to be given in that "calculation"? How many/what proportion of a group must agree for there to be consensus? Does the absence of a consequence to an act or behavior validate that behavior as a norm? What kind of reaction must occur for an act to be defined as or considered deviant? If norms change over time, does "deviance" also change? If so, what drives these changes? And yet additional issues are raised by the definitions of deviance. Deviance, according to some of these definitions, is linked to power (those who define deviance) or its absence (those who are defined as "deviant"). What constitutes deviance is linked to social status. And deviance may exist at the far end of a spectrum of tolerance. However, if classification of a behavior as deviant is linked to social status and power, then can the same behavior be classifiable as "not deviant" based on context and/or observer? If there exists a spectrum of tolerance with respect to individuals' acts or behaviors, is there also a spectrum of tolerance relating to the consequences that can be meted out in response to an act that is termed deviant?

Norm Exaggeration

LeVine (1982) has argued that deviance may result not only from norm or rule violation, but also from norm exaggeration. He posits that a cultural environment may induce individuals to behave in a manner that exaggerates "normal" and culturally distinctive behavior. He argues, for instance, that the relatively high rate of suicide among the Japanese may, in fact, be an exaggeration of the cultural value of self-sacrifice.

The degree to which individuals conform to a norm may vary. LeVine (1982), for instance, developed the idea of the J-curve. At the lower (left) side of the curve is a small percentage of individuals who regularly violate the norm. The center portion of the curve represents the percentage of individuals who have infrequently violated the norm and, for the most part, have conformed to it. The right side of the curve reflects the largest proportion of the population, those who have consistently acted in conformity with the norm.

Exceptions: Nondeviant Rule Violations

Edgerton (1985) explored situations in which rule violations are not deemed to be acts of deviance. He noted, for instance, that some rules have exceptions. Others are only loosely enforced, while still others are enforced with surprising rigidity. How is it that some rule violations will provoke a consequence while others are not enforced? How is it that some rules have exceptions? To some extent, this depends on the nature of the rule, as indicated, and on the status of the violator. However, the consequences may also turn on the account of the violation that is proffered by the violator (Edgerton, 1976). That account, or explanation, can exacerbate the violator's situation, excuse it, or justify it (Edgerton, 1976; Scott & Lyman, 1968). The process of account giving and consequence "is usually more akin to a negotiation than to a mechanical or impartial effort to support propriety or administer justice" (Edgerton, 1976: 30).

Under the normative approach, then, behaviors or statuses that are outside the boundaries or push the boundaries of what is considered "normal" by consensus will be considered deviant and subject to consequences. This could include, for instance, numerous behaviors such as the use of alcohol (MacAndrew & Edgerton, 1969), violent behavior (Edgerton, 1976), and some forms of addictive behavior, such as illicit drug use or smoking. Deviance, according to this approach, also encompasses departures from an "ideal" or normative status, so that those who look different or speak differently are to be considered deviant and, potentially, suspect. We see this theme repeated across numerous groups that are perceived as engaging in unsavory behavior, such as those who "can't hold their drink."

The Social Audience Approach

The "social audience" approach defines deviance quite differently, with an emphasis on those viewing the behavior rather than on a violation of a norm:

> Basically the ultimate measure of whether or not an act is deviant depends on how others who are socially significant in power and influence define the act.... One could commit any act, but it is not deviant in its social consequences if no elements of society react to it (Bell, 1971: 11).

> [A]cts and actors violating the norms of society will be termed "rule-breaking behavior" and "rule breakers," while the terms "deviant behavior" and "deviant" will be reserved for acts and actors labeled as deviant by a social audience (Cullen and Cullen, 1978: 8).

> Deviance is not a property *inherent* in certain forms of behavior; it is a property *conferred upon* these forms by the audience which directly or indirectly witnesses them (Erikson, 1962: 308).

Lemert (1951) has explained how this dynamic comes about. He notes that norms vary across time and place and that members of society become aware of norms only after they have been violated. Such deviation can be measured objectively:

> [T]he sum total of deviation in a given situation will consist of the variance of the actions from the prescribed social norms multiplied by the number of persons who engage in such actions (Lemert, 1951: 51).

The societal response to this deviation corresponds to the degree and visibility of the deviation. In some cases, the response is extreme in proportion to the actual breach. Lemert (1951: 56) explains such "spurious" reactions as the result of "rivalry or conflict of groups ... as they aspire to power or struggle to maintain their position."

Lemert distinguished between primary deviation, which refers to instances of norm violation in which individuals do not view themselves as deviant, and secondary deviation. Secondary deviance arises from the societal response to primary deviance:

> When a person begins to employ his deviant behavior or a role based upon it as a means of defense, attack, or adjustment to the ... problems created by the consequent societal reaction to him, his deviation is secondary (Lemert, 1951: 76).

The Audience Role

Becker (1963) expanded on Lemert's theory by identifying distinct aspects of the social audience's role: (1) the formulation of rules; (2) the application of these rules; and (3) the labeling of particular individuals as outsiders or deviants. The process through which individuals are so labeled often depends more on who is doing the labeling and who is being labeled than on the nature of the behavior involved. Consequently, behavior cannot be labeled as deviant until after the response has occurred. Becker (1963: 9) explained:

> Social groups create deviance by making the rules whose infraction constitutes deviance, and by applying those rules to particular people and labeling them as "outsiders." From this point of view, deviance is not a quality of the act the person commits, but rather a consequence of the application by others of rules and sanctions to an offender. The deviant is one to whom that label has successfully been applied; deviant behavior is that people so label.

Kitsuse (1962) posited that audiences create deviance through a three-stage process: (1) interpretation of behavior as deviant; (2) definition of individuals who engage in that behavior as deviant; and (3) response to the individual(s) in a manner considered appropriate to the specific type of deviance. An individual is not considered deviant until the third stage, when he or she is subject to the verbal or physical repercussions of the behavior.

Ben-Yehuda (1990: 36) has argued that "who interprets whose behavior, why, where, and when is very crucial ... ." He offers two specific examples, those of Joan of Arc and of the Nobel Laureate chemist Louis Pauling. Joan of Arc, convicted of being a heretical deviant and executed in 1431, was later canonized by Pope Benedict XV. Pauling, supported by thousands of scientists from all over the world, lobbied the United Nations in the late 1950s to end nuclear weapons testing. His efforts led to an interrogation by the U.S. Senate and he was prohibited from attending various international scientific meetings. In 1962, he was awarded the Nobel Prize for his efforts in promoting peace (Ben-Yehuda, 1990).

Like the normative approach, however, the "social audience" definition is also unable to address all issues satisfactorily. Kitsuse's (1962: 253) definition of deviance underscores some of the difficulties with this approach:

> Forms of behavior per se do not differentiate deviants from non-deviants; it is the responses of the conventional and conforming members of the society who identify and interpret behavior as deviant which sociologically transform persons into deviants.

It is unclear, however, what kind of response, both in nature and degree, is either necessary or sufficient to merit classification of a particular behavior as deviant. It is also unclear what kind of acts, both in nature and degree, are necessary or sufficient to provoke a necessary and sufficient response. If there is no response to certain acts, such as murder and torture, can it be said that these acts are not "deviant?" But, if these acts are inherently "deviant," regardless of audience response, can it really be said then that classification as "deviant" rests entirely on definition by the audience, without reference to a norm?

Responses to Deviance

A variety of responses are available to both the "deviant" individual and to those witnessing the "deviance." Responses of the "deviant" may include the proffering of an account, secondary deviance, and/or attempts to cover deviance.

Responses of the "Deviant"

The Account

Scott and Lyman (1968: 46) have defined an "account" as "a linguistic device employed whenever an action is subjected to valuative inquiry." They elaborate:

> By an account, then, we mean a statement made by a social actor to explain unanticipated or untoward behavior—whether that behavior is his own or that of others, and whether the proximate cause for that statement arises from the actor himself or from someone else. An account is not called for when people engage in

> routine, common-sense behavior in a cultural environment that recognizes that behavior as such.

Scott and Lyman (1968) distinguish between accounts that are excuses and those that constitute justifications. Excuses will mitigate or relieve the actor from his or her responsibility. Excuses include appeals to accidents, to defeasibility, to biological drives, and to scapegoating. Justifications will not only neutralize an act or its consequences, but will also assert the positive value of the act. Justifications include denial of injury, denial of the victim, condemnation of the condemners, and appeal to loyalties. Whether any account will be honored will depend on the character of the social circle in which it is utilized and the expectancies of the audience receiving the account.

The response of the audience may well depend on the nature of the account that is offered. For instance, where the audience response is likely to be one of deterrence, *i.e.*, punishment, the account given to explain a criminal act may determine (a) whether there will, in fact, be any punishment and (b) the extent of that punishment, if imposed. Drunken driving provides one example of where an account may be critical to the response.

Secondary Deviance

Secondary deviance arises from the societal response to primary deviance, which refers to instances of norm violation in which individuals do not view themselves as deviant:

> When a person begins to employ his deviant behavior or a role based upon it as a means of defense, attack or adjustment to the ... problems created by the consequent societal reaction to him, his deviation is secondary (Lemert, 1951: 76).

Covering

Covering refers to attempts to hide the deviance from others, in an attempt to "blend" with the nondeviant (French, Wilke, Mayfield, and Woolley, 1985). Covering attempts may include the use of a wig to hide baldness due to chemotherapy or elaborate routines to "cover" one's mental retardation. Goffman (1963) documented, for instance, one blind man's attempts to hide his blindness from a date. Throughout numerous dates and several movies, this gentleman held the hand of his date, so that she unwittingly led him where he needed to go, giving her the perception that he was a sighted person. Edgerton (1993) documented in detail the efforts of formerly institutionalized mentally retarded adults to "pass" by lying about where they were from to avoid discovery of the prior institutionalization; by marrying to appear normal and to emphasize their status as a free person; by acquiring "memorabilia" so as to have a history; by finding excuses to explain their lack of a car; and by devising mechanisms to simplify others' responses, such as asking whether it is 9:00 yet (responses: almost, no, just after) rather than asking the exact time (responses: 8:40, 20 minutes to 9, etc.). Covering, or "passing," has been used by members of communities of color to gain access to privileges and benefits once accorded to only whites and by gays to avoid social and economic sanctions for being homosexual (Chauncey, 1994; Lopez, 1994; Polednak, 1997). Individuals who abuse alcohol, for instance, may attempt to cover by hiding the extent of their drinking.

Audience Response

Gove (1982) has delineated various classes of audience response to deviance: social reaction, deterrence, prevention, incentives, treatment, amelioration, and self-curative. To these can be added acceptance (Bogdan & Taylor, 1987). The literature relating to each of these strategies is extensive and will only be touched upon briefly.

Social Reaction

Social reaction advocates the redefinition of a situation so that the behavior initially viewed as a problem is no longer viewed as such. Some might argue, for instance, that the establishment of a needle exchange program, in essence, redefines illicit drug use by providing the mechanism by which it can be accomplished, thereby redefining the situation so that illicit drug use is no longer perceived as a problem.

Punishment/Deterrence

Punishment or deterrence is one of the most common strategies for dealing with deviant behavior. The use of a deterrence approach is premised on operant theory, which posits that individuals perform an act when the rewards of that act exceed the costs (Gove, 1982). Operant theory suggests, then, that punishment, as in the criminal justice system, must be systematically and rapidly applied. Increased efforts to address drunken driving are illustrative of this approach (Robertson, 1997). Neither, however, may occur, due to a variety of reasons, including overburdening of the courts and plea bargaining (Gove, 1982).

Prevention

Prevention focuses on the identification of what is perceived to be the weak link in a complex causal process and the manipulation of that link in order to prevent the potentially developing deviance. This strategy, commonly used in the health context, encompasses a broad range of interventions, including legislation, such as gun control measures (Gove, 1982) and tobacco control (Cummings, 1997), education (Gove, 1982), and screening (Champion & Miller, 1997).

Incentives

Gove (1982) notes that the use of incentives to reward nondeviant behavior, thereby encouraging nondeviant behavior while simultaneously discouraging deviant behavior, is rarely utilized. One example is the use of employee incentive programs to encourage attendance and discourage absences due to illness.

Treatment

Treatment is one of the most common responses to deviance and often occurs in response to deviance forms that have become medicalized, such as alcoholism or drug addiction. The form of the treatment necessarily depends on the form of deviance or malady.

Self-Cure

Gove (1982) indicates that some forms of deviance are believed to be self-curative, e.g., an individual will "grow out of it."

Acceptance

Bogdan and Taylor (1987: 35) define acceptance as a

> [relationship] ... between a person with a deviant attribute and another person, which is of long duration and characterized by closeness and affection and in which the deviant attribute does not have a stigmatizing, or morally discrediting, character. Accepting relationships are not based on a denial of difference, but rather on the absence of impugning the different person's moral character because of the variation.

Such relationships are generally founded on feelings of family, a sense of religious obligation or commitment, humanitarian orientation, and feelings of friendship. Accepting relationships are often characterized by formation in stages over a period of time; increasing levels of trust and comfort as the relationship develops; shared contempt between the "deviant" and the nondeviant for "outsiders" who manifest their discomfort; de-emphasis of the differences and emphasis on the positive aspects of the relationship; and empathy from the nondeviant individual with the discrimination and/or rejection experienced by the "deviant."

DEVIANCE AT THE JUNCTURE OF EPIDEMIOLOGY AND LAW

As Pearce (1996) has indicated, epidemiology does not operate in a societal vacuum. Indeed, out notion of what should be studied is inextricably linked to our notion of what constitutes disease or a departure from the desired state of affairs. How we then explain that disease, or the absence of the disease, may similarly reflect not only the state of our knowledge at the time, but also our preexisting beliefs about what is deviant or out of the range of the ordinary.

Consider, for example, the initial portrayal of the HIV epidemic by epidemiologists. On June 26, 1982, a Haitian refugee who had been detained by the Immigration and Naturalization Service (INS) at the Krome Detention

Facility in Florida died of toxoplasmosis. She and other Haitians who were found to be infected with HIV denied both homosexual activity and injection drug use (Farmer, 1992). In Haiti itself, 61 cases of AIDS were identified between June 1979 and October 1982. Those infected included men and women, heterosexuals, bisexuals, and homosexuals.

For want of a better explanation, U.S. physicians and other scientists inferred an association between being Haitian and being HIV-infected. One government physician informed the public that these cases could reflect "an epidemic Haitian virus that was brought back to the homosexual population in the United States" (Chabner, 1982), while another attributed the infection to voodoo practices (Moses & Moses, 1983). Respectable medical journals perpetuated the myth that HIV somehow originated through Haitian voodoo practices:

> Even now, many Haitians are voodoo *serviteurs* and partake in its rituals ... (Some are also members of secret societies such as Bizango or "impure" sects, called "cabrit thomazo," which are suspected to use human blood itself in sacrificial worship. As the HTLV-III/LAV virus is known to be stable in aqueous solution at room temperature for at least a week, lay Haitian voodooists may be unsuspectingly infected with AIDS by ingestion, inhalation, or dermal contact with contaminated ritual substances, as well as by sexual activity (Greenfield, 1986: 2200).

This portrayal of Haitians as a disease-bearing population was further enlarged even more with the February 1990 prohibition by the Food and Drug Administration of blood donations from individuals of Haitian ethnicity. Haitians, together with others designated as members of "risk groups" (homosexuals, heroin users, and hemophiliacs), became, as a result of less than thoughtful pronouncement from health professionals, including epidemiologists (Centers for Disease Control, 1983), the deviant. Membership in one of these groups came to be synonymous with HIV infection, much as it had been associated already with different values and different lifestyles. In designating these groups as "high risk," the Centers for Disease Control (CDC) essentially ignored the potential adverse consequences that could befall these groups' members socially, economically, and politically (Oppenheimer, 1988; Nachman & Dreyuss, 1986) and failed to realize the ultimate harm that such a stance would pose to objective scientific inquiry:

> A ... consequence of creating high-risk groups was to reinforce the relationship between the disease and "marginal" members of the population. In the case of HIV, although each of the groups ostensibly threatened the remainder of the community through the medium of blood or sex, public health recommendations would inhibit such contamination. Consequently, the disorder could be contained at the boundaries, among people who were "different" from the majority but undifferentiated within each of these high-risk groups.
>
> One of the dangers of a scientific classification of people based on stereotypes was that it defined the questions raised and thus answered (Oppenheimer, 1992: 62).

Unfortunately, this chapter of HIV epidemic illustrates to well the adverse consequences that may result from less that thorough epidemiological investigations and from careless and inaccurate pronouncements from the media.

Many Haitians had attempted to immigrate to the United States, driven from Haiti by excruciating poverty and political violence at the hands of the Duvalier regime and its henchmen. The INS, however, often characterized many of those fleeing not as political refugees, which would have permitted them to enter the United States in search of refuge, but as economic refugees, which entitled them to nothing more than a return ticket home if they actually entered.

A group of Haitian refugees was interdicted and held for processing at the United States military base at Guantanamo Bay (*Haitian Centers Council v. Sale*, 1993). Those who were found to be HIV-positive were not admitted into the United States as refugees, although after screening for political asylum it was found that they did, indeed, meet the legal standard for admission to the U.S. as refugees. Rather, they were held in a separate part of the detention camp, known as Camp Bulkeley. These 200 refugees were held as prisoners, although they had not committed any crimes. They were denied access to the camps facilities. They received medical care from military physicians, who did not have access to the technology necessary to treat adequately individuals suffering from AIDS. The military itself had requested that a number of the detainees be evacuated from the camp in order to receive more adequate medical care, but the INS denied this request each time that it was made. Duane "Duke" Austin, then INS Special Assistant to the Director of Congressional and Public Affairs, remarked that "[T]hey're going to die anyway, aren't they?"

A lawsuit was filed against the INS on behalf of the Haitian refugees. The lawsuit claimed that the practices of INS violated numerous constitutional provisions due to the indefinite length of detention, the use of this detention process that did not comport with any of the procedures that then-existed, and the denial of critically needed medical care. The federal district court that heard the case found that "The Haitian camp at Guantanamo is the only known refugee camp in the world composed entirely of HIV positive refugees" (*Haitian Centers Council v. Sale*, 1993: 1045). The court enjoined the continued detention of the Haitian HIV-seropositive refugees and ordered their release to any location other than Haiti.

DISCUSSION QUESTIONS

1. In addition to the exclusion of Haitians for HIV disease, there have been numerous other instances in our history of excluding individuals of a specified ethnicity or nationality based upon a putative association with certain diseases. For instance, we severely limited the ability of Chinese to enter, believing that they were carriers of leprosy, and of Irish persons, based upon a belief that they suffered from a greater proclivity to mental illness. During the early 1900s we imposed quarantines in the Philippines in an effort to protect Americans from the Philippine "filth" that had produced cholera

(Hayes, 1998). In each of these named situations, moral degeneracy and/or differentness were implicated as a cause of the disease, even by health professionals.

a. Discuss the ethical obligations of epidemiologists with respect to those who are afflicted with a particular disease, as well as their obligations to the larger, general community. How does the involvement of epidemiologists in such pronouncement comport, or fail to comport, to the guidelines enunciated by the Council for International Organizations of Medical Sciences (1991), together with the World Health Organization (1993)?
b. Are incidents such as those mentioned above in (1) reflective of the normative approach, labeling theory, or both? Justify your response. How is the classification of the response relevant to attempts to prevent such misunderstandings in the future, as they relate to disease and to populations?
c. How can epidemiologists today avoid repeating such errors in the future? What strategies can be employed in disseminating research findings while still avoiding the stigmatization of the study population and the community from which it is drawn?

2. Consider the movement of Mothers Against Drunk Driving. To what extent did this movement succeed because of changes in public perceptions of alcohol use, drunkenness, and drunken driving as deviant or not deviant? In evaluating the societal changes in the disposition of drunken driving events, to what extent has each of the following factors played a role in bringing about such changes: (1) changes in the public's perceptions of deviance, (2) epidemiological findings, (3) the political power of consumer/interest groups, and (4) legislative and regulatory changes?

REFERENCES

Becker, H. S. (1973). *Outsiders: Studies in the Sociology of Deviance.* New York: Free Press.

Bell, R. B. (1971). *Social Deviance.* Homewood, Illinois: Dorsey.

Ben-Yehuda, N. (1990). *The Politics and Morality of Deviance: Moral Panics, Drug Abuse, Deviant Science, and Diverse Stigmatization.* Albany, New York: State University of New York Press.

Bierstedt, R. (1963). *The Social Order*, 2nd ed. New York: McGraw-Hill.

Bogdan, R., & Taylor, S. (1987). Toward a sociology of acceptance: The other side of the study of deviance. *Social Policy, 18*, 34–39.

Centers for Disease Control. (1983). Acquired immunodeficiency syndrome update. *Morbidity and Mortality Weekly Report, 32*, 465–467.

Chabner, B. (1982). Cited in *Miami News*, Dec. 2, at 8A.

Champion, V. L., & Miller, A. (1997). Adherence to mammography and breast self-examination regimens. In D. S. Gochman (Ed.). *Handbook of Health Behavior Research II: Provider Determinants* (pp. 245–267). New York: Plenum Press.

Chauncey, G. (1994). *Gay New York: Gender, Urban Culture, and the Making of the Gay World, 1890–1940.* New York: Basic Books.

Clinard, M. B., & Meier, R. F. (1979). *Sociology of Deviant Behavior*, 5th ed. New York: Holt, Rinehart, and Winston.

Council for International Organizations of Medical Sciences (CIOMS). (1991). *International Guidelines for Ethical Review of Epidemiological Studies.* Geneva: CIOMS.

Council for International Organizations of Medical Sciences (CIOMS), World Health Organization (EHO). (1993). *International Guidelines for Biomedical Research Involving Human Subjects.* Geneva: World Health Organization.

Cullen, F. T., & Cullen, J. B. (1978). *Toward a Paradigm of Labeling Theory.* Lincoln, Nebraska: University of Nebraska.

Cummings, K. M. (1997). Health policy and smoking and tobacco use. In D. S. Gochman (Ed.). *Handbook of Health Behavior Research IV: Relevance for Professionals and Issues for the Future* (pp. 231–251). New York: Plenum Press.

DeFleur, M. L., Antonio, W. V., DeFleur, L. B., Nelson, L., & Adamic, C. H. (1977). *Sociology: Human Society*, 2nd ed. Glenview, Illinois: Scott, Foresman.

Dohrenwend, B. P. (1959). Egoism, altruism, anomie and fatalism: A conceptual analysis of Durkheim's types. *American Sociological Review, 24*, 466–473.

Edgerton, R. B. (1993). *The Cloak of Competence*, rev. ed. Berkeley, California: University of California Press.

Edgerton, R. B. (1976). *Deviance: A Cross-Cultural Perspective.* Menlo Park, California: Benjamin Cummings Publishing Company.

Edgerton, R. B. (1985). *Rules, Exceptions, and Social Order.* Berkeley, California: University of California Press.

Erikson, K. T. (1962). Notes on the sociology of deviance. *Social Problems, 9*, 307–314.

Farmer, P. (192). *AIDS and Accusation: Haiti and the Geography of Blame.* Berkeley, California: University of California Press.

French, F., Wilke, A. S., Mayfield, L., & Woolley, B. (1985). The physician's role in covering deviance: Assisting the physically handicapped. *Psychological Reports, 57*, 1255–1259.

Goffman, E. (1963). *Stigma: Notes on the Management of Spoiled Identity.* New York: Simon and Schuster, Inc.

Gove, W. R. (1982). The formal reshaping of deviance. In M. M. Rosenberg, R. A. Stebbins, A. Turowetz eds. *The Sociology of Deviance* (pp. 175–201). New York: St. Martin's Press.

Greenfield, W. (1986). Night of the living dead II: Slow virus encephalopathies and AIDS: Do necromantic zombiists transmit HTLV-III/LAV during voodooistic rituals? *Journal of the American Medical Association, 256*, 2199–2200.

Gutierrez, G. (1987). *On Job: God-Talk and the Suffering of the Innocent.* Maryknoll, New York: Orbis Books.

Hays, J. N. (1998). *The Burdens of Disease: Epidemics and Human Response in Western History.* New Brunswick, New Jersey: Rutgers University Press.

Haitian Centers Council v. Sale. (1993). 823 F. Supp. 1028 (E.D.N.Y.).

Jasanoff, S. (1993). What judges should know about the sociology of science. *Judicature, 77*, 77–82.

Kitsuse, J. I. (1962). Societal reaction to deviant behavior: Problems of theory and method. *Social Problems, 9*, 247–256.

Lemert, E. M. (1951). *Social Pathology.* New York: McGraw Hill.

Lopez, I. F. H. (1994). The social construction of race: Some observations on illusion, fabrication, and choice. *Harvard Civil Rights—Civil Liberties Law Review, 29*, 1–62.

MacAndrew, C., & Edgerton, R. B. (1969). *Drunken Comportment: A Social Explanation.* Chicago: Aldine Publishing Company.

Merton, R. K. (1966). Social problems and sociological theory. In R. K. Merton, R. Nisbet (Eds.) *Contemporary Social Problems*, 2nd ed (pp. 775–823). New York: Harcourt, Brace, and World.

Nachman, S., & Dreyfuss, G. (1986). Haitians and AIDS in South Florida. *Medical Anthropology Quarterly, 17*, 32–33.

Oppenheimer, G. M. (1992). Causes, cases, and cohorts: The role of epidemiology in the historical construction of AIDS. In E. Fee, D. M. Fox (Eds.). *AIDS: The Making of a Chronic Disease* (pp. 49–83). Berkeley, California: University of California Press.

Oppenheimer, G. M. (1988). In the eye of the storm: The epidemiological construction of AIDS. In E. Fee, D. Fox (Eds.). *AIDS: The Burdens of History* (pp. 267–300). Berkeley, California: University of California Press.
Orcutt, J. D. (1983). *Analyzing Deviance*. Homewood, Illinois: Dorsey Press.
Packard, R. (1989). *White Plague, Black Labor: Tuberculosis and the Political Economy of Health and Disease in South Africa*. Berkeley, California: University of California Press.
Pearce, N. (1996). Traditional epidemiology, modern epidemiology, and public health. *American Journal of Public Health, 86*, 678–683.
Moses, P., & Moses, J. (1983). Haiti and the acquired immune deficiency syndrome. *Annals of Internal Medicine, 99*,565.
Polednak, A. P. (1997). *Segregation, Poverty, and Mortality in Urban African Americans*. New York: Oxford University Press.
Reiss, I. (1986). *Journey Into Sexuality*. Englewood Cliffs, New Jersey: Prentice-Hall.
Roberston, L. S. (1997). Health policy, health behavior, and injury control. In D. S. Gochman (Ed.). *Handbook of Health Behavior Research IV: Relevance for Professionals and Issues for the Future* (pp. 215–230). New York: Plenum Press.
Scott, M. B., & Lyman, S. M. (1968). Accounts. *American Sociological Review, 35*, 46–62.
Simmons, J. L. (1965). Public stereotypes of deviants. *Social Problems, 13*, 223–232.
Simon, W. (1994). Deviance as history: The future of perversion. *Archives of Sexual Behavior, 23*, 1–20.
Thio, A. (1978). *Deviant Behavior*. Boston: Houghton Mifflin.
Williams, R. M. Jr. (1968). The concept of norms. In D. L. Sills, ed. *International Encyclopedia of the Social Sciences*. New York: Macmillan, 11: 204–208.
Williams, W. L. (1992). *The Spirit and the Flesh: Sexual Diversity in American Indian Culture*. Boston: Beacon Press.
Young, J. (1977). The police as amplifiers of deviancy. In P. Rock (Ed.). *Drugs and Politics* (pp. 99–134). New Brunswick, New Jersey: Transaction Books.

CHAPTER 11

Case Study Seven

Sex, Gender, and Sexuality

Simple distinctions come all too easily. Frequently we open the way for later puzzlement by restricting the options we take to be available. So, for example, in contrasting science and religion, we often operate with a simple pair of categories. On one side there is science, proof, and certainty; on the other, religion, conjecture, and faith ... (Kitcher, 1982).

DEFINING SEX

It is generally assumed that sex and gender are sexually dimorphic, meaning that there exists a phylogenetically inherited structure of two types, male and female (Herdt, 1994). Lillie's (1939) thoughts represent an early explanation of such sexual dimorphism:

> What exists in nature is a dimorphism within species into male and female individuals, which differ with respect to contrasting characters, for each of which in any given species we recognize a male form and a female form, whether these characters be classed as of the biological, or psychological, or social orders. Sex is not a force that produces these contrasts; it is merely a name for our total impression of the differences ... In the strictly historical sense of these words, a male is to be defined as an individual that produces spermatozoa; a female one that produces ova; or individuals at least having the characters associated with these functions.

The identification of an individual as a biological male or female rests on an evaluation of one or more of the following elements: chromosomal sex, gonadal sex, morphological sex and secondary sex traits, and psychosocial sex or gender identity (Herdt, 1994). In general, it has been assumed that biological sex, determined by one or more of the first three elements, is inextricably linked with gender identity, as well as with gender role/social identity and sexual orientation (Bolin, 1994). This underlying assumption has been subject to ever-growing dispute in recent years. Each of these elements is discussed in greater detail below.

This general overview of the development of biological sex presumes a basic knowledge of anatomy and human reproduction. Individuals wishing further details are urged to consult Moore and Persaud (1993).

The chromosomal sex of an embryo depends on whether the fertilization of the ovum occurs by an X-bearing or Y-bearing sperm. Fertilization by an X-bearing sperm results in an XX zygote, which will normally develop into a female. Fertilization by a Y-bearing sperm will produce an XY zygote, which normally develops into a male (Moore & Persaud, 1993). However, the gonads of both males and females are identical in appearance prior to the seventh week of the embryo's development; during this period, they are referred to as indifferent or undifferentiated gonads.

Hermaphroditism or intersexuality occurs when there appears to be a discrepancy between the morphology of the gonads (testes or ovaries) and the appearance of the external genitalia. (For a discussion of the development of the testes and ovaries, see Moore and Persaud, 1993). True hermaphroditism is extremely rare and occurs only when both testicular and ovarian tissue are present. However, these tissues are generally nonfunctional (Krob, Braun, & Kuhnle, 1994; Moore & Persaud, 1993; Talerman, Verp, Senekjian, Gilewski, & Vogelzang, 1990).

GENDER IDENTITY AND GENDER ROLE

Stoller (1968: viii–ix) distinguished between sex as a function of biology and gender as a function of culture:

> Dictionaries stress that the major connotation of sex is a biological one as, for example, in the phrases *sexual relations* or *the male sex* ... It is for some of these psychological phenomena [behavior, feelings, thoughts, fantasies] that the term *gender* will be used: one can speak of the male sex or the female sex, but one can also talk about masculinity and femininity and not necessarily be implying anything about anatomy or physiology.

Gender has been defined as

> a multidimensional category of personhood encompassing a distinct pattern of social and cultural differences. Gender categories often draw on perceptions of anatomical and physiological differences between bodies, but those perceptions are always mediated by cultural categories and meanings .... Gender categories are not only "models of" difference...but also "models for" difference. They convey gender-specific social expectations for behavior and temperament, sexuality, kinship and interpersonal roles, occupation, religious roles and other social patterns. Gender categories are "total social phenomena" ... a wide range of institutions and beliefs find simultaneous expression through them, a characteristic that distinguishes gender from other social statuses (Roscoe, 1994: 341).

Despite this distinction, (biological) sex has often been considered synonymous with or predictive of gender (social role). The following excerpt illustrates how biological sex was determinative of social function:

> Thus it was claimed that women's low brain weights and deficient brain structures were analogous to those of lower races, and their inferior intellectualities explained on this basis. Woman, it was observed, shared with Negroes a narrow,

> childlike, and delicate skull, so different from the more robust and rounded heads characteristic of males of 'superior' races. Similarly, women of higher races tended to have slightly protruding jaws, analogous to, if not as exaggerated as, the apelike jutting jaws of lower races. *Women and lower races were called innately impulsive, emotional, imitative rather than original, and incapable of the abstract reasoning found in white men* (Stepan, 1990: 39–40) (emphasis added).

Gender identity and gender role are distinct. Nanda's (1994: 395–396) explanation is instructive:

> Gender identity has been defined as the private experience of gender role: the experience of one's sameness, unity and the persistence of one's individuality as male, female, or androgynous, expressed in both self-awareness and in behavior. Gender role is everything that a person says and does to indicate to others or to the self the degree to which one is either male, female or androgynous. Gender role would thus include public presentations of self in dress and verbal and nonverbal communication; the economic and family roles one plays; the sexual feelings (desires) one has and the persons to whom such feelings are directed; the sexual role one plays and emotions one experiences and displays; and the experiencing of one's body, as it is defined as masculine or feminine in any particular society. Gender identity and gender role are said to have a unity, like two sides of a coin.

Stoller's conceptualization of gender identity and gender role similarly distinguish between the private and the public experiences:

> I am using the word *identity* to mean one's own awareness (whether one is conscious of it or not) of one's existence or purpose in this world or, put a bit differently, the organization of those psychic components that are to preserve one's awareness of existing (Stoller, 1968: x).

Gender identity is to be further distinguished from "core gender identity," a "person's unquestioning certainty that he belongs to one of only two sexes" (Stoller, 1968: 39). Stoller elaborated as follows:

> This essentially unalterable core of gender identity [I am a male] is to be distinguished from the related but different belief, *I am manly* (or masculine). The latter attitude is a more subtle and complicated development. It emerges only after the child has learned how his parents expect him to express masculinity (Stoller, 1968: 40).

This, too, can be contrasted with gender role. "Core gender identity" signifies the feeling of "I am a male" or "I am a female," whereas "gender role" represents "a masculine or feminine way of behaving" (Walinder, 1967: 4). Gender identity is to be distinguished, as well, from sexual identity:

> The term "gender identity" [is] used ... rather than various other terms which have been employed in this regard, such as the term "sexual identity." "Sexual identity" is ambiguous, since it may refer to one's sexual activities or fantasies, etc .... Thus, of a patient who says "I am not a very masculine man," it is possible to say that his gender identity is male although he recognizes his lack of so-called masculinity (Stoller, 1964: 220).

SEXUAL ORIENTATION

We have come to equate the choice of one's sexual partner—male or female—with one's sexual orientation. It is clear, though, that homosexual behavior is not synonymous with homosexuality. For instance, a recent study of male sexual behavior in the United States found that 2% of the respondents aged 20 to 39 reported having had any same sex sexual activity during the previous 10 years, but only 1 percent reported exclusively same sex sexual activity for the same time period (Billy, Tanfer, Grady, & Keplinger, 1993). Identical sexual acts, including the choice of the sexual partner, may vary in meaning and significance, depending on their cultural and historical context (Vance, 1995). As illustrated by the following examples taken from both history and current events, we see that the biological sex of one's sexual partner may have as much or more to do with power, economic relations, and the availability of alternative partners as with sexual desire or orientation.

In a number of societies, sexual relations between younger and older men were structured by age (Greenberg, 1988). The older male often assumed the active role in the relationship, while the younger male assumed the passive role. The sexual act could include masturbation, anal intercourse, and/or fellatio. The motivation for these relationships varied depending on the culture, but could include the sexual transmission of special healing powers from the older male to the younger disciple; the belief that a boy must have semen implanted in his body by an adult in order to mature physically; a fear that heterosexual intercourse would deplete men's vitality; a scarcity of women; and a belief that heterosexual intercourse would harm men because of women's polluting qualities (Greenberg, 1988).

Homosexual behavior—as distinct from homosexuality—may also reflect a differential in power between the sexual partners. This includes, for example, the use of one's body rather than money to deter violence, a situation that may arise both on the street (Scacco, 1982), in prisons, and during war (Greenberg, 1988; Trexler, 1995).

The status of berdache and hijra both exemplify relations that might be termed homosexual based on only a superficial understanding of sexual behavior but, in fact, signify something quite different. Williams (1992: 2) defined a berdache as "a morphological male who does not fit society's standard man's role, who has a nonmasculine character." Native Americans often referred to berdaches as "halfmen-halfwomen," although they were neither hermaphrodites nor transsexuals (Williams, 1992). Berdaches, who are now known as "two-spirit people" (Lang, 1996), existed within various Native American tribes, including the Cheyenne, Creek, Klamath, Mohave, Navaho, Pima, Sioux, and Zuni (Greenberg, 1988; Roscoe, 1991; Williams, 1992). Rather than being termed effeminate, two-spirit people are more accurately described as androgynous. They are not perceived as either men or women, but rather as an alternative gender. Often, two-spirit persons combined the behavior, social roles, and dress of both men and women. Although some assumed the passive role in a sexual relationship with another man, the sexual relationship was a secondary component of status as a berdache (Callender & Kochems, 1986; Williams, 1992). Homosexual behavior was not synonymous with status as a berdache (Williams, 1992).

Similarly, some "two-spirit" women adopted some male roles and dress and had sexual relations with women (Schaeffer, 1965). As a result of missionary efforts and United States government agents, the berdache tradition has declined. Younger Native Americans may have rejected the role of the berdache in favor of a self-identity as a gay male (Williams, 1992).

The hijras of India have been characterized as "neither man nor woman and woman and man" (Nanda, 1990; Nanda, 1994). Based on that status, hijras play a religious role, derived from Hinduism, by blessing newborn male children and performing at marriages (Nanda, 1990). Hijras are defined as such by their lack of desire for and sexual impotence with women, rather than by their sexual relations with men. Their impotence with women is attributed to a defective or absent male sexual organ, which is lacking due to an accident of birth or to intentional emasculation, the surgical removal of the male genitals (Nanda, 1990). In defining themselves, hijras collapse sex and gender into one category, defining themselves as "not men" because of their impotence with women and as "not women" because of their inability to bear children. They simultaneously incorporate aspects of the female role, such as dress, gendered erotic fantasies, desire for male sexual partners, and a gender identity of a woman or hijra, with those of a male role, such as coarse speech and the use of the hookah for smoking (Nanda, 1994). Clearly, despite their sexual relations with men, hijras are not defined and do not define themselves as homosexuals.

Homosexuality has been variously conceived of as an innate, relatively stable condition (Murray, 1987a); a congenital, but not hereditary condition (Heller, 1981); a form of congenital degeneracy (Gindorf, 1977); an earlier evolutionary form of the human race, i.e., bisexual or hermaphroditic (Krafft-Ebing, 1965); a perverse and immature orientation resulting from family interactions during childhood development (Dynes, 1987); and the result of psychological processes similar to those that lead to heterosexuality, modifiable through various forms of therapy (Akers, 1977). It was not until 1973 that the American Psychiatric Association removed homosexuality as a mental illness from the *Diagnostic and Statistical Manual of Psychiatric Disorders* (Greenberg, 1988).

TRANSSEXUALITY AND TRANSGENDERISM

Transsexuality

Transsexuals have been defined as "individuals with a cross-sex identity," regardless of their surgical status (Bolin, 1992: 14), while transsexuality has been classified as a gender identity disorder resulting in "clinically significant distress or impairment in social, occupational, or other important areas of functioning" (Reid & Wise, 1995: 241). Whether transsexuality per se constitutes a disorder, which then necessitates treatment, is subject to considerable dispute (Loue, 1996). Diagnosis rests on the presence of "a strong and persistent cross-gender identification" and "a persistent discomfort with one's sex or a sense of inappropriateness in the gender role of that sex (Reid & Wise, 1995: 241). Diagnosis requires that the condition be distinguished from hermaphroditism, a desire to change sex in order to gain a cultural or social advantage, and a desire to change sex due to nonconformity with prescribed sex roles (Reid & Wise, 1995: 240).

Researchers have estimated that approximately 1 out of every 11,900 men (male to female) and approximately 1 out of every 30,400 women (female to male) are transsexuals (Bakker, van Kesteren, Gooren, & Bezemer, 1993). Estimates of the sex ratio have varied widely, from 2.5 men to 1 woman in the Netherlands to 5.5 women to 1 man in Poland (Bakker et al., 1993; Godlewski, 1988; Pauley, 1968).

Treatment for transsexualism has included long-term hormonal therapy and sex change surgery. Genital reassignment surgery from female to male is complex and extensive and, consequently, must generally be performed in several stages (Hage, Bouman, de Graaf, & Bloem, 1993). Phalloplasty is used to construct a penis in a female-to-male transsexual (Hage, Bloem, & Suliman, 1993). In addition to the surgery, female-to-male transsexuals must often adhere to a long-term regimen of androgen administration (Sapino, Pietribiasi, Godano, & Busolati, 1992). Potential adverse effects include necrosis, hernia, venous congestion, and phallic shaft fistulas (Hage, Bloem, & Suliman, 1993). Male-to-female transsexuals may also undergo extensive surgery (Eldh, 1993) and hormonal treatment (Valenta, Elias, & Domurat, 1992). Potential adverse effects include the lack of a sensate clitoris (Eldh, 1993), vaginal stenosis (Crichton, 1992; Stein, Tiefer, & Melman, 1990), and pain during sexual intercourse (Stein, Tiefer, & Melman, 1990). Many transsexuals may ultimately decide to forgo the surgery due to its high cost, the lack of health insurance coverage for the surgery (Gordon, 1991; Stein, Tiefer, & Melman, 1990), and the fear of an unsatisfactory surgical outcome (Crichton, 1992; Hage, Bout, Bloem, & Megens, 1993).

Transgenderism

The word "transgender" has several meanings. First, it can encompass all those who "challenge the boundaries of sex and gender" (Feinberg, 1996: x). Alternatively, it refers to those who reassign the sex they were labeled at birth, and those whose expressed gender is considered inappropriate for their apparent sex (Feinberg, 1996). A distinction is often made between transsexuals, who change or attempt to change the sex that they were assigned at birth, and transgender individuals, who "blur the boundar[ies] of the *gender expression*" traditionally associated with the biological sexes (Feinberg, 1996: x).

Such blurring may take the form of cross-dressing (Garber, 1992). Cross-dressing, i.e., wearing the clothing that is most often associated with the opposite sex, may occur for various reasons and in numerous contexts.

Women may assume an "imitation man look" in an attempt to be more successful in business (Molloy, 1977). Men, some homosexual and some heterosexuals, may perform as female impersonators. Gay men may dress in drag as a means of self-assertion or activism (Garber, 1992). Cross-dressing has been central in theater (Baker, 1968; Heriot, 1975) and, to some degree, in religion (Barrett, 1931; Garber, 1992; Warner, 1982). It is important to recognize that not all of these are associated with being a transgendered individual.

HOMOSEXUALITY, SCIENCE, AND CHANGING SOCIAL CONTEXT

Homosexuality in Science and Medicine: Changing Views

In the United States, homosexuality has been understood until relatively recently to be an abnormal, or deviant, medical condition deserving of some form of treatment. These views derive from both a succession of teachings from those deemed to be experts on such matters, as well as an exaggeration or misunderstanding of others. Freud's work, for instance, has been widely and wrongly interpreted as providing evidence that homosexuality itself was evidence of a mental problem. The following excerpts, however, indicate that although Freud believed that homosexuality was not an advantage, neither did he believe that it was a condition necessitating treatment or evidencing disturbance in and of itself (Mondimore, 1996).

> It is the high esteem felt by the homosexual (as a young child) for the male organ which decides his fate. In his childhood he chooses women as his sexual object, so long as he assumes that they too possess what is in his eyes is an indispensable part of the body (a penis); when he becomes convinced that women have deceived him in this particular, they cease to be acceptable to him as a sexual object. He cannot forgo a penis in anyone who is to attract him to sexual intercourse; and if circumstances are favorable he will fix his libido upon the "woman with a penis," a youth of feminine appearance (Freud, 1909).
>
> The explanation [for a girl's love of an older woman] is as follows. It was just when the girl was experiencing the revival of her infantile Oedipus complex at puberty that she suffered her great disappointment. She became keenly conscious of the wish to have a child, and a male one; that what she desired was her father's child and an image of him, her consciousness was not allowed to know. And what happened next? It was not she who bore the child, but her unconsciously hated rival, her mother. Furiously resentful and embittered, she turned away from her father and from men altogether. After her first great reverse she forswore her womanhood and sought another goal for her libido (Freud, 1920).
>
> I gather from your letter that your son is a homosexual. I am most impressed by the fact that you do not mention this term yourself in your information about him ... Homosexuality is assuredly no advantage, but it is nothing to be ashamed of, no vice, no degradation, it cannot be classified as an illness; we consider it to be a variation of the sexual function produced by a certain arrest in development. Many highly respectable individuals of ancient and modern times have been homosexuals, several of the greatest men among them ... It is a great injustice to

> persecute homosexuality as a crime and cruelty to ... What analysis can do for your son runs in a different line. If he is unhappy, neurotic, torn by conflicts, inhibited in social life, analysis may bring him harmony, peace of mind, full efficiency, whether he remains homosexual or gets changed (Freud, 1935).

Homosexuality, Science, and Civil Rights

Homosexuality, Science, and Medicine

Sexual Behavior

It was not until the publication of Kinsey's research on sexual behavior, however, that the notion of homosexuality as a vice or a disease was challenged. Kinsey, a biologist by training, embarked on his research on human sexuality so as to better understand the subject matter of the course he was to teach.

Kinsey discovered in the course of his research that it was not possible to categorize individuals as either heterosexual or homosexual. Accordingly, he and his colleagues developed the 7-point Kinsey Heterosexual-Homosexual Rating Scale. This scale measured both physical contact and psychological reaction along a spectrum ranging from 0 to 6: 0 corresponded to exclusive heterosexuality; 1, signified predominately heterosexual, only incidentally homosexual; to 6, exclusively homosexual (Kinsey, Pomeroy, & Martin, 1948). Kinsey found that 37 percent of the men surveyed had had at least one same-sex sexual experience to the point of orgasm at sometime in their lives, while 10 percent of the men reported being exclusively homosexual for a period of at least 3 years. The Kinsey research group expressed surprise at these findings:

> Whether the histories were taken in one large city or another, ... in small towns, or in rural areas, whether they came from one college or from another, a church school or a state university or some private institution, whether they came from one part of the country or from another, the incidence data on the homosexual have been more or les the same .... Persons with homosexual histories are to be found in every age group, in every social level, in every conceivable occupation, in cities and on farms, and in the most remote areas of the country (Kinsey, Pomeroy, & Martin, 1948: 625).

Kinsey and colleagues later commented on the apparent unwillingness of both professionals and laypersons to accept his findings:

> It is characteristic of the human mind that it tries to dichotomize in its classification of phenomena. Things are either so, or they are not so. Sexual behavior is either normal or abnormal, socially acceptable or unacceptable, heterosexual or homosexual; and many persons do not want to believe that there are gradations in these matters from one to the other extreme (Kinsey, Pomeroy, Martin, & Gebhard, 1953: 469).

Later work by Hooker focused on the analysis of psychological test results that had been administered to both heterosexual and homosexual men. On the basis of her findings, Hooker concluded that homosexuality does not exist as a clinical entity, that homosexuality as a sexual pattern falls within the normal psychological range, and that the role of specific forms of sexual desire and expression in personality and development was less than had been previously believed (Hooker, 1957).

Medicine, however, largely ignored these investigations for decades. It was not until 1976 that the American Psychiatric Association removed homosexuality from its publication *Diagnostic and Statistical Manual of Mental Disorders*, the publication relied on by clinicians in many industrialized countries, including the United States, to diagnose mental disorders. One must, then, necessarily ask what prompted the American Psychiatric Association to modify its position. That answer can be found in the increasing body of genetic research related to homosexuality and the changing social and political backdrop against which those investigations were conducted.

Genetics and a Role for Epidemiology

Most recently, investigations into the biological causes and attributes of homosexuality have focused on brain function (Hall and Kimura, 1995; McCormick and Witelson, 1994; Gladue and Beatty, 1990), brain structure (Scamyougeras, Witelson, Bronskill, Stanchev, Black, Cheung, Steiner, & Buck, 1994; Allen & Gorski, 1992), and sexual genetics (Hu, Pattatucci, Patterson, Fulker, Cherny, Kruglyak, and Hamer, 1995; Hamer, Hu, Magnuson, Hu, and Pattatucci, 1993; Macke, Hu, Hu, Bailey, King, Brown, Hamer, & Nathans, 1993). Genetic epidemiology may be of particular relevance here, as epidemiological methods applied to the field of genetics may provide insight into the genetic causes of specified attributes. What these genetic studies have revealed to date is that heredity may account for part, but only part, of the development of sexual orientation.

Other research has focused on the development of gender non-conforming behaviors, such as a boy's affinity for dolls, or a girl's "tomboyish" behavior. These nonconforming interests and behaviors often began during childhood, lending additional support to the theory that sexual orientation is part of the "hard wiring" in each individual (see Isay, 1990).

Homosexuality and a Changing Context

"Coming out" as homosexual or lesbian was rendered extremely difficult prior to the 1970s due to the stigma associated with such identity. Individuals were perceived as being deviant, in need of treatment. Those who were identified as homosexual risked losing their jobs, their families and, essentially, everything that they held dear. One gay man explained that homosexuals

> are a minority, not only numerically, but also as a result of a caste-like status in society .... Our minority status is similar, in a variety of respects, to that of national, religious, and other ethnic groups: in the denial of civil liberties; in the legal, extra-legal and quasi-legal discrimination; in the assignment of an inferior social position; in the exclusion from the mainstream of life and culture (Cory, 1951: 13–14).

The hostility suffered by homosexuals and lesbians was exemplified by their treatment in government. Then-Senator Joseph McCarthy, joined by other conservative politicians, demanded the expulsion of homosexuals from the ranks of government employees. The Senate Committee on Expenditures in Executive Departments (1950) warned that the presence of even a single "sex pervert in a Government agency" could

> have a corrosive influence upon his fellow employees. These perverts will frequently attempt to entice normal individuals to engage in perverted practices. This is particularly true in the case of young and impressionable people who might come under the influence of a pervert .... One homosexual can pollute a Government office.

Not surprisingly, the military accelerated its identification and exclusion from its ranks of those believed to be homosexual. The characterization of homosexuals and lesbians as moral perverts and risks to national security emboldened police forces to conduct raids on places believed to be gathering sites for gays. External, unrelated events, such as mayoral elections or the reported slaying of a young boy, could easily provoke an intensification of such raids (D'Emilio, 1983).

The movement for gay rights began quietly, with the formation of the Mattachine Society in southern California in 1950. What began as a discussion group evolved into a membership society that was involved in the defense of those entrapped by police practices, one that published a gay magazine. Eventually, however, the membership of some of Mattachine's members in the Communist Party became a liability for the organization, during an era when McCarthy was waging his campaign against communists. And, despite the formation of this group,

> [t]he claim that homosexuals were a minority with a distinctive culture was still too much at odds with the situation of gay men and women. The dominant view of homosexuality, with its emphasis on the individual nature of the phenomenon, more accurately described gay existence (D'Emilio, 1983: 91).

Gradually, however, other gay-focused organizations were formed. A body of gay nonfiction developed. Gay characters were included in films, although the manner in which they were depicted was often inaccurate and derogatory (D'Emilio, 1983). Journalists writing about homosexuals were sometimes sympathetic, but sometimes expressed a view that can only be described as vitriolic:

> [I]t is a pathetic little second-rate substitute for reality, a pitiable flight from life. As such it deserves fairness, compassion, understanding and, when possible, treatment. But it deserves no encouragement, no glamorization, no rationalization, no fake status as a minority martyrdom, no sophistry about simple differences in taste—and, above all, no pretense that it is anything but a pernicious sickness (Editorial, Jan. 21, 1966: 41).

The increasing visibility of homosexuals and homosexual life prompted questions within the psychoanalytic profession as to the real causes of homosexuality. Some professionals were willing to reexamine, and discard, what had become the traditional notions of causality and cure. These skeptics were joined by their colleagues in sociology, who

> mounted the sharpest attack on traditional interpretations of homosexuality. The investigation of deviance had a long history .... The postwar era witnessed a significant shift, however, as a radical relativist perspective crept into the area. Social scientists came to view deviance not as evidence of social disorganization but, rather, as a sign of different norms of behavior. They examined deviant populations as members of a subculture, or they made use of labeling theory which emphasized the functions that the attribution of deviant status to a group served for the majority (D'Emilio, 1983: 142).

One such sociologist, applying this perspective to the "problem" of homosexuality, concluded:

> The best solution to the problem of homosexuality is one which is modeled on the solution to the problem of religious difference, namely, a radical tolerance for homosexual object-choice, whether as a segment of an individual's sexual existence or as a full commitment to homosexuality as a way of life .... Society as a whole must significantly shift its attitudes towards homosexuals and homosexuality in American life (Hoffman, 1968: 197–198).

The 1960s witnessed the assassinations of John Kennedy and Martin Luther King, civil rights demonstrations, protests across college campuses, and protests against the war in Vietnam. Against this backdrop, significant legal strides were made for those who identified as gay, including the repeal of some sodomy laws. In San Francisco, an openly gay man ran for political office and a gay community center was opened. The recognition by a group of liberal clergy of the unjust police practices directed against the gay community, and the clergy's ensuing political activism on behalf of the gay community, signaled a watershed in the visibility and tolerance of the gay community within San Francisco.

Not surprisingly, a new militant element emerged in the gay community, largely attributable to Franklin Kameny (Clendinen & Nagourney, 1999; D'Emilio, 1983). Kameny urged the gay community to model its efforts on the strategies utilized by the civil rights movement. Kameny rejected attempts to explain how or why homosexuality existed, arguing that such inquiry would be akin to asking which gene was responsible for the color of a black person's skin. In contrast to the attempts by many gay "activists" to accommodate the concerns of "mainstream" communities, Kameny asserted that homosexual acts engaged in by consenting adults represented a positive behavior for the individuals themselves and for society (D'Emilio, 1983). Kameny, and others, persuaded the American Civil Liberties Union to defend the civil rights of gays, which included challenging police practices of harassment, the imposition of criminal penalties for sodomy performed between consenting adults, and discrimination in employment. These more militant members also attempted to persuade others in the gay community to reject the medical model of homosexuality. (For a discussion of the hostility that arose between the militant members of the movement and the more conservative members, see D'Emilio, 1983.)

On June 27, 1969, the New York City police conducted what was originally planned as a routine raid on a gay bar, the Stonewall Inn. They had raided the bar many times before, without protest from the bar's patrons. On this particular occasion, however, the crowd fought back. Thus was born the Gay Liberation movement, situated against a backdrop consisting of the black civil rights movement, protests against the Vietnam war, and a women's rights movement. The following night, Christopher Street bore graffiti proclaiming "Gay Power!" Protests against the disparate treatment of homosexual continued: the Gay Activist Alliance disrupted the presidential campaign of John Lindsay in New York, by handcuffing themselves to railings. The Gay Liberation Front declared,

> We are a revolutionary group of men and women formed with the realization that complete sexual liberation for all people cannot come about unless existing social institutions are abolished. We reject society's attempt to impose sexual rules and definitions of our nature. We are stepping outside these roles and simplistic myths. We are going to be who we are. At the same time, we are creating new social forms and relations, that is, relations based upon brotherhood, cooperation, human love, and uninhibited sexuality. Babylon has forced us to commit ourselves to one thing—revolution! (Gay Liberation Front Statement of Purpose, 1969).

Numerous protests prompted the American Psychiatric Association to examine the scientific work of Kinsey and Hooker. Ultimately, the general membership, as well as an appointed scientific committee of the Association, voted to delete the listing of homosexuality as a psychiatric illness in the *Diagnostic and Statistical Manual of Mental Disorders.*

DISCUSSION QUESTIONS

1. Much of the groundbreaking research into the causes of homosexuality and the characteristics of homosexuals has been conducted by biologists, geneticists, psychologists, and sociologists.
   a. Explain the difference, if any, between the role to be played by behavioral epidemiologists, as compared with sociologists and psychologists, and by genetic epidemiologists, as compared with geneticists.
   b. How have researchers limited the scope and nature of the research questions posed by postulating the existence of only two biological sexes (male and female), in lieu of accepting multiple categories, such as male, female, and intersex? How would a change in this underlying assumption affect the nature or the scope of research relating to sexual orientation?
2. Significant opposition to the depiction of homosexuals and homosexuality as non-deviant remains, as evidenced by laws that ban gay marriages, that continue to criminalize sodomy between consenting adults, that prohibit the adoption of children and/or the legal custody

of natural-born children by their homosexual parents; and that fail to recognize as marriages gay marriages entered into in foreign jurisdictions. To what extent can change effectuated through research be maintained over time? Compare and contrast your assessment with the change that has occurred for other minority communities as the result of the civil rights movement of the 1960s and 1970s.

REFERENCES

Akers, R. L. (1977). *Deviant Behavior: A Social Learning Approach.* Belmont, California: Wadsworth.

Baker, R. (1994). *Drag: A History of Female Impersonation in the Performing Arts.* New York: New York University Press.

Bakker, A., van Kesteren, F. J. M., Gooren, L. J. G., & Bezemer, P. D. (1993). The prevalence of transsexualism in the Netherlands. *Acta Psychiatrica Scandinavia, 87*, 237–238.

Barrett, W. P. (Trans.). (1931). *The Trial of Jeanne d'Arc.* London: Routledge.

Billy, J. O. G., Tanfer, K., Grady, W. R., & Klepinger, D. H. (1993). The sexual behavior of men in the United States. *Family Planning Perspectives, 25*, 52–60.

Bolin, A. (1992). Coming of age among transsexuals. In T. L. Whitehead & B. V. Reid (Eds.) *Gender Constructs and Social Issues* (pp. 13–39). Chicago: University of Chicago Press.

Bolin, A. (1994). Transcending and transgendering: Male-to-female transsexuals, dichotomy and diversity. In G. Herdt (Ed.), *Third Sex, Third Gender: Beyond Sexual Dimorphism in Culture and History* (pp. 447–485). New York: Zone Books.

Callonder, C., & Kochems, L. M. (1986). Men and not-men: Male gender-mixing and homosexuality. *Anthropology and Homosexual Behavior.*

Clendinen, D., & Nagourney, A. (1999). Out for Good: The Struggle to Build a Gay Rights Movement in America. New York: Simon and Schuster.

Cory, D. W. (1951). The Homosexual in America. New York.

Crichton, D. (1992). Gender reassignment surgery for male primary transsexuals. *South African Medical Journal, 83*, 347–349.

D'Emilio, J. (1983). Sexual Politics, Sexual Communities: The Making of a Homosexual Minority in the United States, 1940–1970. Chicago: University of Chicago Press.

Dynes, W. (1987). *Homosexuality: A Research Guide.* New York: Garland.

Eldh, J. (1993). Construction of a neovagina with preservation of the glans penis as a clitoris in male transsexuals. *Plastic Reconstructive Surgery, 91*, 895–900.

Feinberg, L. (1996). *Transgender Warriors.* Boston: Beacon Press.

Freud, S. (1909). Analysis of a phobia in a five year old boy. Standard Edition, 10, 1–147.

Freud, S. (1935). Letter to an American mother, in American Journal of Psychiatry (1951), 107, 786.

Freud, S. (1920). The psychogenesis of a case of homosexuality in a woman. In P. Rieff (Ed.) (1963). *Sexuality and the Psychology of Love* (pp. 133–59). New York: Collier.

Garber, M. (1992). *Vested Interests: Cross-Dressing and Cultural Anxiety.* New York: HarperCollins.

Gay Liberation Front. (1969). Statement of Purpose, July 3. Reprinted in J. D'Emilio (1983). Sexual Politics, Sexual Communities: The Making of a Homosexual Minority in the United States, 1940–1970. Chicago: University of Chicago Press.

Gindorf, R. (1977). Wissenschaftliche Ideologien im Wandel: Die Angst von der Homosexualitat als intellektuelles Ereignis. In J. S. Hohmann (Ed.) *Der underdruckte Sexus* (pp. 129–144). Berlin: Andreas Achenbach Lollar. Cited in D. F. Greenberg (1988). The Construction of Homosexuality. Chicago: University of Chicago Press.

Godlewski, J. (1988). Transsexualism and anatomic sex: Ratio reversal in Poland. *Archives of Sexual Behavior, 17*, 547–548.

Gordon, E. B. (1991). Transsexual healing: Medicaid funding of sex reassignment surgery. *Archives of Sexual Behavior, 20*, 61–79.
Greenberg, D. F. (1988). *The Construction of Homosexuality*. Chicago: University of Chicago Press.
Hage, J. J., Bloem, J. J. A. M., & Suliman, H. M. (1993). Review of the literature on techniques for phalloplasty with emphasis on the applicability in female-to-male transsexuals. *Journal of Urology, 150*, 1093–1098.
Hage, J. J., Bouman, F. G., de Graaf, F. H., & Bloem, J. J. A. M. (1993). Construction of the neophallus in female-to-male transsexuals: The Amsterdam experience. *Journal of Urology, 149*, 1463–1468.
Hage, J. J., Bout, C.A., Bloem, J. J. A. M., & Megens, J. A.J. (1993). Phalloplasty in female-to-male transsexuals: What do our patients ask for? *Annals of Plastic Surgery, 30*, 323–326.
Hamer, D., Hu, S., Magnuson, V., Hu, N., & Pattatucci, A. (1993). A linkage between DNA markers on the X chromosome and male sexual orientation. Science, 261, 321–327.
Heller, P. (1981). A quarrel over bisexuality. In G. Chapple & H. H. Schulte (Eds.) *The Turn of the Century: German Literature and Art, 1890–1915* (pp. 87–115). Bonn: Bouvier Verlag Herbert Grundmann.
Herdt, G. (1994). Third sexes and third genders. In G. Herdt (Ed.), *Third Sex, Third Gender: Beyond Sexual Dimorphism in Culture and History* (pp. 21–81). New York: Zone Books.
Heriot A. (1975). *The Castrati in Opera*. New York: Da Capo Press.
Hoffman, M. (1968). The Gay World: Male Homosexuality and the Social Creation of Evil. New York: Basic Books.
Hooker, E. (1957). The adjustment of the male overt homosexual. Journal of Projective Techniques, 21, 18–31.
Hu, S., Pattatucci, A., Patterson, C., Fulker, D., Cherny, S., Kruglyak, L., & Hamer, D. (1995). Linkage between sexual orientation and chromosome Xq28 in males but not in females. Nature Genetics, 11, 248–256.
Isay, R. (1990). Psychoanalytic theory and the therapy of gay men. In D. McWhirter, S. Sanders, & J. Reinisch (Eds.). Homosexuality/Heterosexuality: Concepts of Sexual Orientation. (New York: Oxford University Press.)
Kinsey, A., Pomeroy, W., & Martin, C. (1948). Sexual Behavior in the Human Male. Philadelphia: W. B. Saunders.
Kinsey, A., Pomeroy, W., Martin, C., & Gebhard, P. (1953). Sexual Behavior in the Human Felame. Philadelphia: W. B. Saunders.
Kitcher, P. (1982). *Abusing Science: The Case Against Creationism*. Boston: MIT Press.
Krafft-Ebing, R. v. (1965). *Psychopathia Sexualis: A Medico-Forensic Study*. Trans. H.E. Wedeck. New York: G. P. Putnam's Sons. Orig. pub. 1886.
Krob, G., Braun, A., & Kuhnle, U. (1994). Hermaphroditism: Geographical distribution, clinical findings, chromosomes and gonadal histology. *European Journal of Pediatrics, 153*, 2–10.
Lang, S. (1996). There is more than just women and men: Gender variance in North American Indian cultures. In S. P. Ramet (Ed.), *Gender Reversals & Gender Cultures: Anthropological and Historical Perspectives* (pp. 183–196). London: Routledge.
Lillie, F. (1939). General biological introduction. In E. Allen (Ed.) *Sex and Internal Secretions: A Survey of Recent Research*, 2nd ed. Baltimore, Maryland: Williams and Wilkins.
Loue, S. (1996). Transsexualism in medicolegal limine: An examination and proposal for change. *Journal of Psychiatry and Law, Spring*, 27–51.
Macke, J., Hu, N., Hu, S., Bailey, M., King, V., Brown, T., Hamer, D., & Nathans, J. (1993). Sequence variation in the androgen receptor gene is not a common determinant of male sexual orientation. American Journal of Genetics, 53, 844–852.
Molloy, J. T. (1977). *The Woman's Dress for Success Book*. New York: Warner Books.
Mondimore, F. M. (1996). A Natural History of Homosexuality. Baltimore, Maryland: Johns Hopkins University Press.
Moore, K. L., & Persaud, T. V. N. (1993). *The Developing Human: Clinically Oriented Embryology*, 5th ed. Philadelphia: W.B. Saunders.
Murray, S. O. (1987a). Homosexual acts and selves in early modern Europe. *Journal of Homosexuality, 15*, 421–439.

Nanda, S. (1994). Hijras: An alternative sex and gender role in India. In G. Herdt (Ed.), *Third Sex, Third Gender: Beyond Sexual Dimorphism in Culture and History* (pp. 373–417). New York: Zone Books.
Nanda, S. (1990). *Neither Man Nor Woman: The Hijras of India.* Belmont, California: Wadsworth Publishing Company.
Pauley, I. B. (1968). The current status of the change of sex operation. *Journal of Nervous and Mental Disease, 147,* 460–471.
Roscoe, W. (1994). How to become a berdache: Toward a unified analysis of gender diversity. In G. Herdt (Ed.), *Third Sex, Third Gender: Beyond Sexual Dimorphism in Culture and History* (pp. 329–372). New York: Zone Books.
Roscoe, W. (1991). *The Zuni Man-Woman.* Albuquerque: University of New Mexico Press.
Sapino, A., Pietribiasi, F., Godano, A., & Bussolati, G. (1992). Effect of long-term administration of androgens on breast tissues of female-to-male transsexuals. *Annals of the New York Academy of Science, 586,* 143–145.
Scacco, A. (Ed.). (1982). *Male Rape: A Casebook of Sexual Aggression.* New York: AMS Press.
Schaeffer, C. E. (1965). The Kutenai female berdache: Courier, guide, prophetess, and warrior. *Ethnohistory, 12,* 193–236.
Senate Committee on Expenditures in Executive Departments, Employment of Homosexuals and Other Sex Perverts in Washington, 81st Cong., 2d Sess. Quoted in J. D'Emilo. (1983). Sexual Politics, Sexual Communities: The Making of a Homosexual Minority in the United States 1940–1970. Chicago: University of Chicago Press.
Stein, M., Tiefer, L., & Melman, A. (1990). Followup observations of operated male-to-female transsexuals. *Journal of Urology, 143,* 1188–1192.
Stepan, N. L. (1990). Race and gender: The role of analogy in science. In D. T. Goldberg (Ed.). *Anatomy of Racism* (pp. 38–57) Minneapolis: University of Minnesota Press.
Stoller, R. J. (1964). A contribution to the study of gender identity. *Journal of the American Medical Association, 45,* 220–226.
Stoller, R. J. (1968). *Sex and Gender: On the Development of Masculinity and Femininity.* New York: Science House.
Talerman, A., Verp, M. S., Senekjian, E., Gilewski, T., & Vogelzang, N. (1990). True hermaphrodite with bilateral ovotestes, bilateral gonadoblastomas and the dysgerminomas, 46, XX/46,XY karotype, and a successful pregnancy. *Cancer, 66,* 2668–2671.
Editorial (1966). Time, Jan. 21, 41.
Trexler, R. C. (1995). *Sex and Conquest: Gendered Violence, Political Order, and the European Conquest of the Americas.* Ithaca, New York: Cornell University Press.
Valenta, L. J., Elias, A. N., & Domurat, E. S. (1992). Hormone pattern in pharmacologically feminized male transsexuals in the California state prison system. *Journal of the American Medical Association, 84,* 241–250.
Vance, C. S. (1995). Social construction theory and sexuality. In M. Berger, B. Wallis, & S. Watson (Eds.). *Constructing Masculinity* (pp. 37–48). New York: Routledge.
Walinder, J. (1967). *Transsexualism: A Study of Forty-Three Cases.* trans. H. Fry. Stockholm: Scandinavian University Books.
Warner, M. (1982). *Joan of Arc: The Image of Female Heroism.* New York: Vintage Books.
Williams, W. L. (1992). *The Spirit and the Flesh: Sexual Diversity in American Indian Culture.* Boston: Beacon Press.

CHAPTER 12

Case Study Eight

The Medical Use of Marijuana

Once you make a habit of Observing, Deducing, and Applying, you may sense a pathway opening up ahead of you ... leading to a deeper understanding of things. You may even feel at times as though you're in some sort of Other Dimension ... But you're not, really; you're just seeing and experiencing Things As They Are, rather than as someone-or-other says they are. And the difference between the two can be considerable (Hoff, 1992: 153–154).

MARIJUANA, ITS USES, AND ITS REGULATION

The Beginnings

Marijuana has been used as a medicinal drug in various cultures for thousands of years (Dixon, 1999). It has been used to treat a variety of ailments, including cramps, lack of appetite, and rheumatism. Its active ingredient, delta-9-tetrahydrocannibol (THC) has been isolated for use in Marinol, which has been available to AIDS and cancer patients to combat nausea and to glaucoma patients to help reduce intraocular pressure (Grey, 1996). Marijuana use does not appear to lead to dependence in most individuals (Perkonigg, Lieb, Hofler, Schuster, Sonntag, & Wittchen, 1999). Additionally, as compared with Marinol, smoked marijuana or marijuana ingested with food is cheaper, acts more quickly and is easier to regulate or adjust in terms of the dosage (Bergstrom, 1997; Satel, 1997).

United States physicians first recognized the potential therapeutic value of marijuana as early as 1840. It was recognized in the 1851 edition of the *United States Dispensatory* as a remedy for numerous disorders, including neuralgia, gout, rheumatism, tetanus, hydrophobia, epidemic cholera, convulsions, chorea, hysteria, mental depression, delirium tremens, insanity, and uterine hemorrhage. Until 1942, it was included in the *United States Pharmacopoeia* as a remedy for poor appetite (Grinspoon & Bakalar, 1997; Grinspoon

& Bakalar, 1995). The eventual adoption of provisions prohibiting and criminalizing marijuana must be understood in the context of the social and political climate prevailing at the time of their inception. Opiates, and especially morphine, were widely used during the Civil War period to alleviate the suffering that accompanied battle-related injuries. The indiscriminate use of these drugs, even for relatively common gastrointestinal complaints, may have led to addiction in many Civil War veterans, resulting in what came to be called "morphinomania" (Bonnie & Whitebread II, 1999). This situation was aggravated by the lack of restrictions placed on druggists who refilled morphine-containing prescriptions. It is believed that the numbers of individuals addicted to opiates was further increased through the indiscriminate sale of medicinal remedies to consumers who were often unaware of their opiate content, resulting in an increasing number of addicts among whites, women, and members of the middle class (Bonnie & Whitebread II, 1999). Concurrently, the use of alcohol and tobacco was increasingly widespread and accepted.

The Initiation of Drug Regulations

The first drug regulations were effectuated by the states and essentially represented a reaction to the individuals perceived to be the primary users of the drugs, rather than the effects of the drugs themselves. Opium use among the Chinese became the focus of state legislation in the late 1800s in Nevada, the Dakotas, California, Montana, Wyoming, Arizona, New Mexico, and Washington. Cocaine use among African Americans in the South was identified as the source of criminal conduct. Ultimately, morphine and cocaine came to be associated with the immoral underworld of pimps, gamblers, and prostitutes. States responded with the enactment of legislation designed to restrict pharmacies' ability to dispense morphine-containing drugs (Bonnie & Whitebread II, 1999). Federal legislation was adopted in 1909 to prohibit the importation of opium other than at designated ports and for any use that was not medicinal in nature (Bonnie & Whitebread II, 1999).

The use of marijuana, though, was unregulated by either the federal government or by any state government until the early 1900s, when the state of California first prohibited its possession or sale (McGuire, 1997) and various other states adopted restrictions on the dispensing of cannabis without a prescription or on the refilling of prescriptions containing the substance (Bonnie & Whitebread II, 1999). The American Medical Association opposed a federal prohibition against the use of marijuana, which was vigorously supported by members of law enforcement. For instance, Eugene Stanley, a district attorney in New Orleans, alleged in 1931 that

> the underworld [had realized] the value of [marijuana] in subjugating the will of human derelicts to that of a master mind. Its use sweeps away all restraint, and its influence may be attributed to many of our present-day crimes. It has been the experience of the Police and Prosecuting Officials in the South that immediately before the commission of many crimes the use of marihuana cigarettes had been indulged in by criminals so as to relieve themselves from the natural restraint

which might deter them from the commission of criminal acts, and to give them the false courage necessary to commit the contemplated crime (Stanley, 1931: 256).

Stanley proposed that federal aid be provided to the states to aid them in their "effort to suppress a traffic as deadly and as destructive to society … ." (Stanley, 1931: 257). Similar assumptions about the effects of the drug are reflected in a report of the same era by the then Surgeon General of the United States:

While the effects of the drug are definitely narcotic in nature, it is habitually taken for the stimulating effects obtained and the individual satisfaction experienced through the temporary inflation of the personality. No evidence exists that the drug is accumulative in its effect or that a tolerance may be developed through its continued use … . The sudden discontinuance of its use … does not give rise to any "withdrawal" symptoms … (Cummings, 1929).

The initial restrictions on the use of marijuana, even for medicinal purposes, stemmed from its classification as a narcotic. That classification rested, in turn, on the association of the drug

with ethnic minorities and with otherwise "immoral" population … . Marihuana was presumed to be addictive, its use inevitably tending to excess. Since its users—Mexicans, West Indians, blacks, and underworld whites—were associated in the public mind with crime, particularly of a violent nature, the association applied also to marihuana, which had a similar reputation in Mexican folklore. (Bonnie & Whitebread II, 1999: 51–52).

This association between the foreign-born and the use and evils of marihuana reflected an already-existing American preoccupation with lawlessness among foreigners (Bonnie & Whitebread II, 1999). The association of marijuana with Mexicans and with crime, in an era of fervent xenophobia, provided ample incentive for the promulgation of restrictive legislation. Eventually, however, law enforcement personnel urged the adoption of restrictions due to the effects of the drug itself, not merely the characteristics of those who were perceived to be users. Marijuana was to become labeled "the most dangerous of all drugs," with the ability to "excite a state of frenzy leading to actions of uncontrollable violence or even murder" (Bonnie & Whitehead II, 1999: 73, citing Hayes and Bowery, 1932).

Prohibitions against the use of marijuana were eventually implemented by the federal government through the enactment of the Marijuana Tax Act of 1937 (Conboy, 2000). That legislation attempted to reduce the use of marijuana through the imposition of a tax. The passage of the Act reflected the triumph of those who had campaigned for regulation on the basis of an alleged association between marijuana use, alienage, crime, and moral depravity, over those who presented a more reasoned and studied response. Bonnie and Whitebread (1999: 153) observed that this "excising of marijuana use from the social organism was seen quite clearly as a means of rooting out idleness and irresponsibility among deviant minorities." This perspective is reflected in an article contained in a 1931 medical journal:

The debasing and baneful influence of hashish and opium is not restricted to individuals but has manifested itself in nations and races as well. The dominant race

> and most enlightened countries are alcoholic, whilst the races and nations addicted to hemp and opium, some of which once attained to heights of culture and civilization have deteriorated both mentally and physically" (Fossier, 1931: 247, quoted in Bonnie & Whitebread II, 1999: 152).

The more reasoned approach is reflected in a report of the Cannabis Subcommittee of the League of Nations Advisory Committee:

> Cannabis indica does not produce a dependence such as in opium addiction. In opium addiction there is a complete dependence and when it is withdrawn there is actual physical pain which is not the case with cannabis. Alcohol more nearly produces the same effect as cannabis in that there is an excitement or general feeling of lifting of personality, followed by a delirious stage, and a subsequent narcosis. There is no dependence or increased tolerance such as in opium addiction. As to the social or moral degradation associated with cannabis it probably belongs in the same category as alcohol. As with alcohol, it may be taken a relatively long time without social or emotional breakdown. Marihuana is habit-forming, although not addicting, in the same sense as alcohol might be with some people, or sugar or coffee. Marihuana produces a delirium with a frenzy which might result in violence; but this is also true of alcohol (League of Nations, 1937).

Following the passage of the Tax Act, the Federal Bureau of Narcotics sought to reduce the level of marihuana-associated sensationalism in the press and to control the cultivation of marijuana for legitimate purposes. This shift in position would render it more difficult for defense attorneys who, prior to the Act's passage, had been able to rely on the government's science-clothed rhetoric to support the insanity defenses offered by their criminal clients (Bonnie & Whitebread II, 1999). Additionally, judges were to be educated to the dangers of marijuana use and the need to impose long-term sentences on would be violators.

The Boggs Act of 1951 sought to discourage further the use of marijuana through the establishment of mandatory prison sentences and monetary fines. These consequences were rendered even more severe with the passage of the 1956 Narcotics Control Act (McGuire, 1997). The subsequent 1970 Controlled Substances Act (CSA) effectively prohibited even the medicinal use of marijuana.

The National Organization for the Reform of Marijuana Laws sought in 1972 to have marijuana reclassified as a Schedule II drug, which would permit its use by prescription. Public hearings on the issue were commenced in 1986 by the Drug Enforcement Agency, which had been previously known as the Bureau of Narcotics and Dangerous Drugs. The DEA's administrative law judge, after hearing extensive testimony from patients and physicians, found in 1988 that the natural form of marijuana was generally therapeutically safe and ordered that the drug be reclassified as a Schedule II substance (Grinspoon & Bakalar, 1995). However, this decision was overridden by the DEA in 1992.

The DEA Administrator relied on the following criteria in rendering his decision: (1) a drug's chemistry must be known and reproducible; (2) there must be adequate safety studies; (3) there must be adequate and well-controlled studies proving efficacy; (4) the drug must be accepted by qualified experts; and (5) the scientific evidence that the drug has a currently accepted medical use must be widely available.

The decision of the DEA was appealed by the Alliance for Cannabis Therapeutics (ACT). ACT argued that marijuana was beneficial in alleviating various side effects of chemotherapy for cancer patients, muscle spasticity in multiple sclerosis patients, and the effects of glaucoma. The court upheld the DEA standards and found that the medicinal use of marijuana failed to comply with these standards. First, there was a dearth of safety studies in humans. Second, well-controlled evaluations of the therapeutic value of marijuana were lacking. Third, the drug's chemistry was unlikely to be known and reproducible due to variations in the plant composition as a function of differences by geographic region in the soil, the light, the water, and harvesting and storage conditions (Margolis, 1994).

Changing Times: States Permit the Medicinal Use of Marijuana

A number of states have recently adopted legislation that has decriminalized the use of marijuana (Institute of Medicine, 1999). And, despite the ongoing existence of federal prohibitions against the use of marijuana, several states have enacted legislation permitting its use in the context of medical treatment (Institute of Medicine, 1999), while in other states, courts have refused to sanction physicians who recommend the medical use of marijuana to their patients (Conant v. McCaffrey, 1999). For instance, California voters passed on November 5, 1996 the Compassionate Use Act of 1996, also known as Proposition 215 (California Health and Safety Code section 11362.5, 1999). The statute permits patients suffering from serious illnesses to possess small quantities of marijuana for their medical use, provides that the primary caregivers of these patients will not be subject to sanctions or criminal prosecution for recommending the use of marijuana, and further protects physicians in California from punishment or loss of rights or privileges for recommending marijuana to a patient for medical use. It is critical to note that the law does not permit physicians to prescribe marijuana to their patients, but permits only a discussion with the patient of the possible benefits and risks, including arrest, of using marijuana for medical purposes. The Clinton administration responded with a threat to prosecute any physician who recommended the use of marijuana, as well as any patients who were found to be using marijuana medicinally (Savage & Warren, 1996).

THE SCIENCE OF MARIJUANA AS MEDICINE

HIV/AIDS and Marijuana

Much of the impetus for the legalization of marijuana in the medical context has come from the suffering of individuals with HIV infection.

A survey of 100 members of the San Francisco Cannabis Cultivator Club, each of whom used marijuana at least once a week, found that 60 of the 100 used the marijuana for the relief of symptoms associated with HIV. Similar use was reported among the members of the Los Angeles buyers club (Institute of Medicine, 1999). Some, but not all, of the individuals using marijuana for the relief of symptoms had previously used the drug recreationally. Why were so many of the clubs' members using marijuana for HIV-related symptoms?

The Epidemiology of HIV/AIDS in the United States

The human immunodeficiency virus, or HIV, is believed to be the causative agent of the disease known as acquired immune deficiency syndrome, or AIDS. The first clinical cases of AIDS were recognized in the United States in 1981, when physicians began to see young gay men infected with *Pneumocystis carinii* pneumonia (PCP) and Kaposi's sarcoma (KS), both of which were highly unusual in young men. These individuals, as well as later-identified hemophiliacs and injecting drug users, shared a critical symptom: immunodeficiency. It was not until 1983, however, that the virus was itself identified. And, although the syndrome AIDS was not recognized as such until the early 1980s, it appears that the disease was present in individuals as early as the 1950s (Osmond, 1999b).

The definition of the syndrome that has come to be known as AIDS has been revised several times since the initial recognition of the disease, in order to accommodate increasing knowledge regarding the effects of the virus. For instance, the initial definition of AIDS did not include HIV wasting and HIV encephalopathy, which were added to the definition of AIDS in 1987; invasive cervical cancer was not included as an AIDS-defining disease until the 1993 revision (Osmond, 1999a).

Initially, the vast majority of identified AIDS cases were attributed to men who had had unprotected sexual intercourse with men, individuals who had shared injection equipment with others, and hemophiliacs and others who had received contaminated blood or blood products. Most recently, an increasing proportion of AIDS cases are occurring among women, among heterosexuals, and among members of communities of color. The greatest proportionate increases since the beginning of the epidemic have occurred in the south and in the Midwest. By 1993, AIDS had become the eighth leading cause of death overall and the leading cause of death among persons aged 25 to 44 (Osmond, 1999b).

Marijuana and the Control of AIDS-Related Symptoms

AIDS-defining illnesses include wasting, defined as "a weight loss of at least 10% in the presence of diarrhea or chronic weakness and documented fever for at least 30 days that is not attributable to a concurrent condition other than human immunodeficiency virus (HIV) infection itself" (Mulligan

& Schambelan, 1999: 403). Prior to the development and adoption of antiretrovirals, the prevalence of wasting was estimated to be as high as 37%. Weight loss also results from nausea from the medication regimen, from candidiasis and other oral manifestations of AIDS that may make eating physically painful, from chronic diarrhea resulting from the illness, from social isolation, and from altered taste perception. Numerous drugs have been utilized in an attempt to stimulate individuals' appetites and/or to reduce the nausea including growth hormone, anabolic steroids, thalidomide, fish oil, and Marinol (Koch, Kim, & Friedman, 1999; Mulligan and Schambelan, 1999).

A number of individuals spoke to members of the Institute of Medicine about the benefits of using marijuana to alleviate both the symptoms of AIDS and the side effects of their medications. One man recounted his many symptoms, which included skin rashes, a metallic aftertaste, dizziness, anemia, headaches, spiking fevers, depression, and neuropathy, as well as uncontrollable nausea, vomiting, and diarrhea. Of marijuana, he said,

> The pot calmed my stomach against a handful of pills. The pot made me hungry so I could eat without a tube. The pot eased the pain of crippling neural side effects so I could dial the phone by myself. The pot calmed my soul and allowed me to accept that I would probably die soon. Because I smoked pot I have lived long enough to see the development of the first truly effective HIV therapies. I lived to gain 50 lb., regain my vigor, and celebrate my 35th birthday (Testimony of G.S., Institute of Medicine, 1999: 27).

A study on the medical use of marijuana of the Institute of Medicine, commissioned by the White House Office of National Drug Control Policy (ONDCP), concluded from its review of animal studies that cannabinoids may, indeed, play a role in the modulation of pain.

Use of Marijuana for Other Disorders

In a study of San Francisco Cannabis Cultivator Club members, it was found that approximately 40 percent of the medicinal marijuana users relied on smoked marijuana to alleviate symptoms of disorders other than HIV/AIDS, including musculoskeletal disorders and arthritis, depression, neurological disorders, gastrointestinal disorders, glaucoma, cancer, and skin manifestations of Reiter's syndrome (Institute of Medicine, 1999). Members of the Los Angeles Cannabis Resource Center reported usage of medicinal marijuana for these and additional disorders, including epilepsy, Tourette syndrome, and multiple sclerosis. Speakers at public workshops conducted by the Institute of Medicine also reported the use of marijuana to treat anorexia, nausea, and vomiting associated with AIDS, cancer, migraines, Wilson's disease, and multiple sclerosis; to treat mood disorders such as depression, anxiety, bipolar disorder, and posttraumatic stress disorder; for the management of pain associated with migraine, injury, postpolio syndrome, degenerative disk disease, rheumatoid arthritis, nail-patella syndrome, the effects of Gulf War chemical exposure, multiple congenital cartilaginous exostosis; to reduce muscle spasticity associated with multiple sclerosis, paralysis, spinal cord injuries, and spasmodic torticollis; to relieve intraocular pressure associated with glaucoma; and to relieve diarrhea associated with Crohn's disease (Institute of Medicine, 1999). An exploratory study conducted by Osborne and colleagues (2000) of a convenience sample of cannabis users found that, in addition to the relief of HIV/AIDS symptoms, marijuana was used to alleviate chronic pain, depression, anxiety, menstrual cramps, and migraine headaches.

TESTING THE EFFECTS OF MARIJUANA

The Institute of Medicine

The Institute of Medicine has issued a series of six recommendations, based on its review of what is currently known about the medical use of marijuana:

1. Research should continue into the physiological effects of synthetic and plant-derived cannabinoids and the natural function of cannabinoids found in the body. Because different cannabinoids appear to have different effects, cannabinoid research should include, but not be restricted to, effects attributable to THC alone.
2. Clinical trials of cannabinoid drugs for symptom management should be conducted with the goal of developing rapid-onset, reliable, and safe delivery systems.
3. Psychological effects of cannabinoids such as anxiety reduction and sedation, which can influence medical benefits, should be evaluated in clinical trials.
4. Studies to define the individual health risks of smoking marijuana should be conducted, particularly among populations in which marijuana use is prevalent.
5. Clinical trials of marijuana use for medical purposes should be conducted under the following limited circumstances: trials should involve only short-term marijuana use (less than six months), should be conducted in patients with conditions for which there is reasonable expectation of efficacy, should be approved by institutional review boards, and should collect data about efficacy.
6. Short-term use of smoked marijuana (less than six months) for patients with debilitating symptoms (such as intractable pain or vomiting) must meet the following conditions:
   - failure of all approved medications to provide relief has been documented,
   - the symptoms can reasonably be expected to be relieved by rapid-onset cannabinoid drugs,

- such treatment is administered under medical supervision in a manner that allows for assessment of treatment effectiveness, and
- involves an oversight strategy comparable to an institutional review board process that could provide guidance within 24 hours of a submission by a physician to provide marijuana to a patient for a specified use

(Institute of Medicine, 1999: 10–11).

FDA Procedures

A review of FDA-required procedures for the conduct of drug-related clinical trials may be helpful here, in view of the recommendations issued by the Institute of Medicine.

Animal Testing

Prior to the initiation of clinical trials, the proposed product must be tested in the laboratory using animals or by testing the product on human tissues in a test tube (in vitro testing). Animal testing is conducted to identify potential toxic effects, to enhance the understanding of the drug's pharmacokinetics, and for data relating to dosage and efficacy, which will be extrapolated to humans.

Investigational New Drug Applications

The sponsor must then submit to the FDA an Investigational New Drug Application (INDA), which sets forth the results of any animal and in vitro testing that have been performed. It is important to note that this requirement applies to the development of all new drugs. In submitting an IND application, the sponsor agrees not to begin clinical investigations until 30 days after the FDA's receipt of the application. If the sponsor is not contacted within 30 days following the FDA's receipt, then it may begin the clinical investigation. Alternatively, the FDA may contact the sponsor in order to obtain more information. In such cases, the initiation of the investigation is delayed pending FDA approval.

There are both commercial and non-commercial INDs. Commercial INDs are those that are submitted in order to obtain marketing approval for a new drug or for the new use of a drug that has already received approval. There are 3 types of non-commercial INDs: the investigator IND, the emergency use IND, and the treatment IND. The investigator IND is submitted by a physician, who both initiates and conducts the investigation. The drug may be given only under the direct supervision of the physician. The emergency IND is utilized to authorize the immediate shipment of experimental drugs in situations that are considered to be urgent. The treatment IND involved drugs that may potentially be used to treat serious or life-threatening conditions. The treatment IND permits these drugs to be used even while clinical work is ongoing and prior to the completion of FDA review.

A drug that is under investigation may not be advertised as safe or effective for the circumstances that are under investigation. Participants in the study may not be charged for the drug without the approval of the FDA.

Clinical Trials

Phase I clinical trials are designed to assess the toxicity of the investigational drug. This phase focuses on an examination of how individuals will metabolize and tolerate the drug. Phase I trials are often conducted with a small number of individuals. Many times, the participants in phase I studies will be healthy volunteers, because the goals of such studies is to assess toxicity, rather than efficacy. Phase I studies are designed to estimate the population maximum tolerated dose (MTD) from the observed individual maximum tolerated doses. The maximum tolerated dose is the lowest quantity at which a single dose of the drug, administered by a specific route, yields an unacceptably high level of toxicity. (For a discussion of how to determine the MTD, see Piantadosi and Liu, 1996; Durieu, Girard, & Boissel, 1990; Storer, 1989). The multiple dose tolerance will be sought after the maximum single dose tolerance is established.

Phase II clinical trials similarly involve a small number of individuals, but here they are individuals who have the disease or condition that is being targeted by the experimental drug. Phase II is designed to assess both the toxicity and the efficacy of the drug. It is not anticipated that the participants in these trials will derive any clinical benefit from their participation.

Phase III clinical trials involve the greatest number of individuals and, depending on the nature of the drug, the disease, and the study design, may require an extended period of time for completion. During this phase, the physicians conducting the drug trials are supervised by the pharmaceutical company that is conducting the test.

New Drug Applications

Following the completion of the clinical trial, the pharmaceutical company seeking approval for the new drug must submit a new drug application (NDA) to the FDA. The NDA must contain all of the results from the animal and human studies that were conducted, information about the proposed uses of the drug, and details about the labeling that will be provided in conjunction with the sale of the drug. When the FDA approves an NDA, the manufacturer can market the drug for the approved indication, but not for off-label uses. Marketing approval is also required if the manufacturer wishes to market a previously approved drug using a new formulation, such as a new mechanism of delivery (inhalation instead of pill form, for example).

The extent to which the testing of the proposed new drug is unbiased remains unclear. Scientists' wish to maintain their professional integrity and the companies' wish to avoid tort liability would mitigate against biased reporting of the testing and reporting of the test findings. However, in some cases, scientists may unconsciously misinterpret data or scientists and/or the pharmaceutical company may consciously fabricate or conceal data (Hillman, Eisenberg, Puly, Bloom, Glick, Kinosian, & Schwartz, 1991; Shapiro, 1978). The General Accounting Office found in one study that the NDA process fails to identify many significant adverse effects that are associated with a new drug (General Accounting Office, 1990). Clearly, then, it may be difficult for the FDA to decide whether or not to approve a new drug, knowing that in some unidentifiable instances, it may approve a drug for use that has adverse effects that outweigh any potential benefit and, in other cases, may delay or deny approval for a drug that is actually beneficial. Notwithstanding the possibility that the FDA may err either on the side of caution or the lack thereof, criticism is more often directed against the FDA for its approval of a new drug:

> For example, in all of FDA's history, I am unable to find a single instance where a Congressional committee investigated the failure of the FDA to approve a new drug. But, the times when hearings have been held to criticize our approval of new drugs have been so frequent that we weren't able to count them ... . The message to FDA staff could not be clearer. Whenever a controversy over a new drug is resolved by its approval, the Agency and the individuals involved likely will be investigated. Whenever such a drug is disapproved, no inquiry will be made. The Congressional pressure for our negative action on new drug applications is, therefore, intense. And it seems to be increasing as everyone is becoming a self-acclaimed expert on carcinogens and drug-testing (Schmidt, 1974, quoted in Grabowski and Vernon, 1983: 5).

Adverse Drug Experience Reports

In order to monitor the effects of a drug more accurately, physicians and consumers may voluntarily file Adverse Drug Experience Reports (ADR) with the pharmaceutical company that markets the drug (21 C.F.R. § 314.80, 2001). The filing of such a report does not mean that the drug caused the adverse effect. Rather, it reflects the fact that the use of the drug and the Observed effect were sufficiently close in time as to suggest that there may be an association between the use and the drug.

Just as there are difficulties associated with the IND process, so are there weaknesses inherent in the ADR procedures. First, in order to file a report, the physician or consumer must believe that there is a connection between the symptoms experienced by the patient and the drug that was prescribed. This may be considerably more difficult if the effect of the drug occurs some time after its use, such as with thalidomide. Second, physicians may be reluctant to file such reports. The reporting is voluntary, and they are under no legal obligation to file them. Many may avoid doing so, believing that to do so will result in their embroilment in potential litigation. Third, the drug manufacturer may not be as scrupulous about filing the required reports with the FDA as would be desired (FDA, 1990; Merrill, 1973). Finally, the FDA's monitoring of such reports may be less than adequate.

The nature of the role to be played by the FDA in the midst of such a labyrinth remains controversial. Some would argue that the FDA should forcefully monitor the activities of the pharmaceutical industry, which is perceived as a threat to the health of consumers as a result of its less than honest disclosures and dealings. Others embrace the concept of the FDA as a partner of the pharmaceutical industry, able to coordinate, as a function of its position, the efforts of the various players involved in the development, manufacture, and marketing of new pharmaceutical products (Tolchin & Tolchin, 1983).

THE POLITICS OF MARIJUANA AS MEDICINE

It is clear, then, that however much public attitudes towards the medicinal use of marijuana may have metamorphosized, significant difficulties attend the conduct of clinical trials designed to assess the efficacy of medicinal marijuana. Politically, the conduct of such research faces an uphill battle. For instance, even though the report of the Institute of Medicine had been commissioned by the "Drug Czar" Barry McCaffrey and the federal Office of Drug Control Policy (Hamilton, 1999), the response of the White House and the Office of Drug Control Policy to the conclusions of that report did little to engender optimism, as illustrated by the following exchange between reporters and White House spokesperson Joe Lockhart:

> Q.: So will the Institute of Medicine's study on medical marijuana influence the administration's position on that subject?
>
> A.: Sure. This—study was done at the request of General McCaffrey and the Office of National Drug Control Policy. I think as General McCaffrey has said out in his news conference that he held a short time ago, that it's important that we have this debate based on science. And this is an important step in the process ... . What we've found out is that there may be some chemical compounds in marijuana that are useful in pain relief or anti-nausea, but that smoking marijuana is a crude delivery system and is not an effective delivery system. So I think what this calls for, as he said, is further research and further research that's like any other research we do in developing drugs through the FDA process, to try to find an effective way to take advantage of the chemical compounds that can be used to fight nausea or pain.
>
> Q.: But beyond the science on this, what about the will of the people in the seven states that have voted to authorize the use of medical marijuana?
>
> A.: Well—
>
> Q.: I mean this White House has often spoken about the will of people—
>
> A.: Right, and I—
>
> Q.: all during the impeachment year.
>
> A.: Right.
>
> Q.: Why is that something that the White House does not respect?
>
> A.: You know, I think we obviously, and as you've pointed out, respect the will of the people. But I think this is a scientific issue, and I'd hate to find—I'd hate

> to see there be a referendum on the latest technology in air traffic control, and I'd hate to see there be a referendum on FDA review process. These are complicated scientific issues, and they ought to be debated on a scientific basis (Federal News Service, 1999).

Shortly later, the "Drug Czar" commented that "there's great potential in the future for this. [The report] also notes, however, that smoked marijuana shows very little promise as medicine" (CNBC, 1999).

A number of ethical issues must also be addressed prior to the initiation of such studies. Previous research has yielded inconsistent findings with respect to the impact of cannabinoids on immune functioning; it is possible that the medicinal use of marijuana among individuals with decreased immune functioning may result in further exacerbation of their condition. Withdrawal symptoms have been observed in animals who are being withdrawn from marijuana. Studies examining the effects of the active ingredients of marijuana on reproductive functioning have similarly yielded inconsistent results. Consequently, our knowledge appears to be inadequate to permit a more complete assessment of both the risks and the benefits of such research.

DISCUSSION QUESTIONS

1. Consider the cases of both tobacco and marijuana. Public smoking is less tolerated presently than it once was, while the use of marijuana—both medical and nonmedical—appears to engender less vociferous opposition than in the past.
   a. To what extent have epidemiological findings driven these changes in attitude?
   b. To what extent have changes in attitude driven epidemiological research relating to tobacco use? To marijuana use?
2. Many potentially addictive drugs are now permitted to be used in the medical context, such as morphine and its derivatives, codeine and its derivatives, amphetamines, and barbiturates. What factors might explain why such substances have been approved for medical use, while marijuana has not received such approval?

REFERENCES

Bergstrom, A. L. (1997). Medical use of marijuana: A look at federal and state responses to California's Compassionate Use Act. *DePaul Journal of Health Care, 2*, 155–182.

Boggs Act, Pub. L. No. 82–235, 65 Stat. 767 (1951), as amended by the Narcotic Control Act of 1956, Pub. L. No. 84–728, 70 Stat. 567 (1956).

Bonnie, R. J., & Whitebread, C. H. II. (1999). *The Marijuana Conviction: A History of Marijuana Prohibition in the United States.* New York: Lindesmith Center.

California Health and Safety Code § 11362.5 (West Supp. 1999).

CNBC. (1999). Rivera Live: Whether marijuana use for medicinal purposes should be legalized. Television broadcast, March 25.

Conboy, J. R. (2000). Smoke screen: America's drug policy and medical marijuana. *Food and Drug Law Journal, 55*, 601–617.

Dixon, J. W. (1999). Case note: *Conant v. McCaffrey*: Physicians, marihuana and the First Amendment. *University of Colorado Law Review, 70*, 975–1017.

Durieu, I., Girard, P., & Boissel, J. P. (1990). Which experimental design for phase I clinical trials? Proposal for an investigational approach. *Fundamentals of Clinical Pharmacology, 4*, 77S–80S.

Federal News Service. (1999). White House Briefing, The White House, Mar. 17. Federal Information Systems Corporation.

Food and Drug Administration (FDA). (1990). New Drug Development in the United States.

Fossier, A. E. (1931). The marihuana menace. *New Orleans Medical and Surgical Journal, 84*. Cited in in R. J. Bonnie, C. H. Whitebread II. (1999). *The Marijuana Conviction: A History of Marijuana Prohibition in the United States*. New York: Lindesmith Center.

Grabowski, H. G., & Vernon, J. M. (1983). *The Regulation of Pharmaceuticals: Balancing the Risks and Benefits*. Washington, D.C.: American Enterprise Institute.

General Accounting Office. (1990). *FDA Drug Review: Postapproval Risks* 1976–85.

Grinspoon, L., & Bakalar, J. B. (1997). *Marihuana, The Forbidden Medicine*. New Haven, Connecticut: Yale University Press.

Grinspoon, L., & Bakalar, J. B. (1995). Marihuana as medicine: A plea for reconsideration. *Journal of the American Medical Association, 273*, 1875–1876.

Hamilton, J. (1999). All things considered: Study released supporting limited medical use of smokable marijuana. National Public Radio broadcast, March 17.

Hayes, M. E., & Bowery, L. E. (1932). Marihuana. *Journal of Criminal Law and Criminology, 23*, 1086–1094. Cited in R. J. Bonnie, & C. H. Whitebread II. (1999). *The Marijuana Conviction: A History of Marijuana Prohibition in the United States*. New York: Lindesmith Center.

Hillman, A. L., Eisenberg, J. M., Pauly, M. V., Bloom, B. S., Glick, H., Kinosian, B., & Schwartz, J. S. (1991). Avoiding bias in the conduct and reporting of cost-effectiveness research sponsored by pharmaceutical companies, *New England Journal of Medicine, 324*, 1362–1365.

Hoff, B. (1992). *The Te of Piglet*. New York: Penguin Group.

Institute of Medicine. (1999). *Marijuana and Medicine: Assessing the Science Base*. Washington, D.C.: National Academy Press.

Koch, J., Kim, L. S., & Friedman, S. A. (1999). Gastrointestinal manifestations of HIV disease. In P. T. Cohen, M. A. Sande, P. A. Volberding (Eds.). *The AIDS Knowledge Base: A Textbook on HIV Disease from the University of California, San Francisco and San Francisco General Hospital* (pp. 523–541). Philadelphia: Lippincott, Williams & Wilkins.

League of Nations. (1937). Advisory Committee on Traffic in Opium and Other Dangerous Drugs, June.

Margolis, R. E. (1994). Marijuana cannot be prescribed for therapeutic purposes. *Healthspan, 3*, 20.

Marijuana Tax Act of 1937, Pub. L. No. 82–235, 65 Stat. 767 (1937), as amended by the Narcotic Control Act of 1956, Pub. L. No. 84–728, 70 Stat. 567.

Merrill, R.A. (1973). Compensation for prescription drug injuries. *Virginia Law Review, 59*, 1–120.

Mulligan, K., & Schambelan, M. (1999). Wasting. In P. T. Cohen, M. A. Sande, & P. A. Volberding (Eds.). *The AIDS Knowledge Base: A Textbook on HIV Disease from the University of California, San Francisco and San Francisco General Hospital* (pp. 403–413). Philadelphia: Lippincott, Williams & Wilkins.

Osborne, A. C., Smart, R. G., Weber, T., & Birchmore-Timney, C. (2000). Who is using cannabis as medicine and why: An exploratory study. *Journal of Psychoactive Drugs, 32*, 435–443.

Osmond, D. H. (1999a). Classification, staging, and surveillance of HIV. In P.T. Cohen, M.A. Sande, P. A. Volberding (Eds.). *The AIDS Knowledge Base: A Textbook on HIV Disease from the University of California, San Francisco and San Francisco General Hospital* (pp. 3–12). Philadelphia: Lippincott, Williams & Wilkins.

Osmond, D. H. (1999b). Epidemiology of HIV/AIDS in the United States. . In P. T. Cohen, M. A. Sande, P. A. Volberding (Eds.). *The AIDS Knowledge Base: A Textbook on HIV*

Disease from the University of California, San Francisco and San Francisco General Hospital (pp. 13–21). Philadelphia: Lippincott, Williams & Wilkins.

Perkonigg, A., Lieb, R., Hofler, M., Schuster, P., Sonntag, H., & Wittchen, H. U. (1999). Patterns of cannabis use, abuse and dependence over time: Incidence, progression, and stability in a sample of 1228 adolescents. *Addiction, 94*, 1663–1678.

Piantadosi, S., & Liu, G. (1996). Improved designs for dose escalation studies using pharmacokinetic measurements. *Statistics in Medicine, 15*, 1605–1618.

Satel, S. (1997). Medical marijuana: Research, don't legalize. *Wall Street Journal*, Oct. 30, A22.

Savage, D. G., & Warren, J. (1996). U.S. threatens penalties if doctors prescribe pot drugs: Criminal charges, other sanctions are possible, officials warn California and Arizona physicians. *Los Angeles Times*, December 31, A3.

Schmidt, A. (1974). The FDA Today: Critics, Congress, and Consumerism. Presentation to the National Press Club, Washington, D.C., October 29. Quoted in H.G. Grabowski, J.M. Vernon. (1983). *The Regulation of Pharmaceuticals: Balancing the Risks and Benefits.* Washington, D.C.: American Enterprise Institute.

Shapiro, S. A. (1978). Divorcing profit motivation from new drug research: A consideration of proposals to provide the FDA with reliable test data. *Duke Law Journal, 1978*, 155–183.

Stanley, E. (1931). Marihuana as a developer of criminals. *American Journal of Police Science, 2*, 256–257.

Storer, B.E. (1989). Design and analysis of phase I clinical trials. *Biometrics, 45*, 935–937.

Tolchin, S., & Tolchin, M. *Dismantling America: The Rush to Deregulate.* Boston: Houghton Mifflin. 21 Code of Federal Regulations § 314.80 (2001).

Index

www.ingramcontent.com/pod-product-compliance
Ingram Content Group UK Ltd.
Pitfield, Milton Keynes, MK11 3LW, UK
UKHW051130260726
13967UKWH00010B/2962

* 9 7 8 1 4 7 5 7 8 7 1 3 9 *